They Play, You Pay

James T. Bennett

They Play, You Pay

Why Taxpayers Build Ballparks, Stadiums, and Arenas for Billionaire Owners and Millionaire Players

Copernicus Books
An Imprint of Springer Science+Business Media

James T. Bennett
Department of Economics
George Mason University
Fairfax, VA, USA
jbennett@gmu.edu

Published in the United States by Copernicus Books,
an imprint of Springer Science+Business Media.

Copernicus Books
Springer Science+Business Media
233 Spring Street
New York, NY 10013
www.springer.com.

Library of Congress Control Number: 2012934954

Manufactured in the United States of America.
Printed on acid-free paper

ISBN 978-1-4614-3331-6 e-ISBN 978-1-4614-3332-3
DOI 10.1007/978-1-4614-3332-3

Acknowledgments

I am grateful to many for their assistance with and support of the research and editing of this book. The research would not have been possible without the generous financial support of the Sunmark Foundation. Research assistance was provided by Homa Saleh. I also owe profuse thanks to my editor, Bill Kauffman, for I am indebted to him for significant contributions to this study.

Fairfax, VA, USA James T. Bennett

Contents

Chapter I

Introduction

If there is one economic truth upon which almost every practitioner of the art or science of economics agrees, it is that publicly financed ballparks, stadiums, and arenas built by taxpayers for professional baseball, football, basketball, and hockey teams are not good investments. We will explore the reasons why in the pages that follow, but one curious fact intrudes itself on any discussion of this matter: Joe Q. Public, instead of being outraged by these rip-offs, is of two minds when it comes to state subvention of professional sports.

On the one hand, he yells and screams about "corporate welfare." Why should he or she, the hard-working taxpaying citizen, have money taken from his or her paycheck and used to build palaces of play for pampered millionaire athletes and arrogant billionaire owners who feel somehow entitled to subsidies reaching into the several hundreds of millions of dollars? Where is the fairness in *that*?

Yet on his other hand — or both hands, since we all know the sound of one hand clapping — he claps and hoots and makes a ruckus cheering on the home team — which really isn't much of a home team, since all or almost all of its members are from elsewhere, live elsewhere in the off-season, and spend most of their fat paychecks elsewhere. But Joe Q. Public, when he isn't grousing as Joe Q. Taxpayer, is Joe Q. Fan. He wears the apparel and logo of his team, whether the Bears or Yankees or Colts or Bills. He watches the team faithfully on television, shouting his approval at the rectangular set or cursing the errors and fumbles of the butterfingered local nine or eleven or six or five.

J.T. Bennett, *They Play, You Pay: Why Taxpayers Build Ballparks, Stadiums, and Arenas for Billionaire Owners and Millionaire Players*, DOI 10.1007/978-1-4614-3332-3_1,

If he can afford the tickets — for they are expensive, and made even more so by the inflationary effect of the subsidized stadium, ballpark, or arena in which the team plays — he will go in person and cheer them on. Parking is ten bucks, and even though the city built the parking lot, the team keeps most or all of the fees. Once inside, he'll pay five dollars for a microwaved hot dog and seven dollars for a lukewarm beer. The consultants who drew up fanciful reports advocating large subsidies for the team probably claimed that fans would spend their pocket money patronizing neighborhood bars and restaurants surrounding the venue, but in recent years savvy owners have realized that by spending a bit more upfront (billed to the taxpayers, most likely), stadium operators can capture the hot dog and beer (and sushi and blooming onion) trade themselves. The proles who can't afford a ticket can watch the game on TV and drink themselves silly at O'Reilly's Tavern across the street, but those watching the game in person will spend their disposable income within the friendly confines of these taxpayer-built edifices.

Believe it or not, the percentage of sports venues that are subsidized by taxpayers has actually declined from its peak in the Big Government heyday of the 1960s and 1970s. But since stadiums are so much more elaborate these days, with their luxury boxes and drive-in-movie-size jumbotrons and fancy club seating and specialty restaurants, the cost to taxpayers continues to rise. And like the snow in that famous Christmas song, it doesn't show signs of stopping.

Once upon a time, ballparks were actually built by the owners of the teams that played in them. Charles Ebbets and Jacob Ruppert and those other owners in the Golden Age may not always have been model citizens, but they didn't beg governments for playpens. They went and had the things built at their own expense, on land they bought from its rightful owners, not stole indirectly via eminent domain. In fact, the most controversial franchise relocation in the history of sport — the movement of baseball's Dodgers from Brooklyn to Los Angeles — involved, as we shall see, one team that played in two storied stadiums, each of which came to be built in a way that was more honorable than dishonorable.

This is not to say, of course, that organized sports in America once existed in an atmosphere of pure laissez-faire. From the beginning, baseball owners used political connections to secure advantages, and as early as 1905 the President of the United States played political football as vigorously as any politician in the succeeding one hundred years. And the gridiron is where our story begins....

Chapter 2

Politics Takes the Field

Although taxpayer-funded ballparks and stadiums are the focus of this analysis, we start with several backstories which illustrate the ways in which government — even the federal government — became involved in sport. After all, hotel taxes to finance football coliseums did not just spring full blown from the brows of greedy owners and their enablers in a hundred local Chambers of Commerce. Government intervention in matters of play grew over time. If the Founding Fathers did not exactly contemplate a role for the federal city in games played with balls and bats, later politicians, whose motivations ran from simple fandom to the cynical pursuit of the pork barrel, entered the field wielding legislation, appropriations, and sometimes threats. And who better to headline our first episode than that apostle of the vigorous outdoor life, President Theodore Roosevelt?

TR Tackles Football

Roosevelt's unprecedented interference in college football "represents a melding of sports and politics that has now become commonplace," as John Sayle Watterson wrote in *The Games Presidents Play: Sports and the Presidency*.[1]

The problems in college football at the turn of the century were legion: unfair play, too many injuries, mercenary athletes, payment to and the deceptive participation of "amateur" players, and poor sportsmanship. (The more things change...) Offense and defense were massed like armies, in phalanxes; the "battering ram" was more than a metaphor, it was a way of moving

J.T. Bennett, *They Play, You Pay: Why Taxpayers Build Ballparks, Stadiums, and Arenas for Billionaire Owners and Millionaire Players*, DOI 10.1007/978-1-4614-3332-3_2,

(and stopping the movement of) the ball. The Intercollegiate Football Rules Committee had a stranglehold over the game, or at least its rules, and efforts at reform annually came to naught. Whether or not the sport was really in danger of abolition has been something of a controversy among sports historians, but certain influential colleges and universities did withdraw from the game, and others threatened to. In addition, Midwestern schools chafed at the effective dictatorship of the East as embodied in the elite-school-dominated rules committee. The Midwesterners wanted to open the game up, make it less clotted and congested and productive of injuries. They wanted "long runs, spectacular plays, and more touchdowns."[2] (Don't we all!) But the East turned a deaf ear to these provincial upstarts.

Until 1905, that is, when President Theodore Roosevelt, who was not terribly bothered by constitutional and historical limits upon presidential power, intervened — and the rules changed. As Guy M. Lewis of the University of Massachusetts-Amherst wrote in his study of "Theodore Roosevelt's role in the 1905 football controversy," TR was "guided by two values: a belief that football was a valuable educational experience for participants and a desire for Harvard teams to be successful." Which was the more powerful motivation is an open question.

That latter motivation is somewhat charming from a distance of a century-plus, but one really must wonder whether or not the Founders who conceived the office of the president and then demarcated its powers had in mind the promotion of the interests of one college's football team over others. Roosevelt "frequently sent the team messages of encouragement," writes Lewis, which is fine, but pressuring a sport's governing body to change the rules might be going just a tad too far, constitutionally.[3] The president's interest was more than just sentiment for the old alma mater. TR's son Kermit played for Groton and his son Ted, though weighing under 150 pounds, lined up for the Harvard freshmen. Wife Edith worried about them, as mothers will do, and she may have played a role in her husband's unprecedented intervention to change the rules of a sport.

The president began his campaign from an accustomed place: the bully pulpit. Speaking on the "Functions of a Great University" at the Alumni Dinner of Harvard on June 28, 1905, President Roosevelt declared:

> I believe heartily in sport. I believe in outdoor games, and I do not mind in the least that they are rough games, or that those who take part in them are occasionally injured. I have no sympathy whatsoever with the overwrought sentimentality which would keep a young man in cotton wool, and I have a hearty contempt for him if he counts a broken arm or collar bone as of serious consequence, when

> balanced against the chance of showing that he possesses hardihood, physical address, and courage. But when these injuries are inflicted by others, either wantonly or of set design, we are confronted by the question not of damage to one man's body, but of damage to the other man's character. Brutality in playing a game should awaken the heartiest and most plainly shown contempt for the player guilty of it, especially if this brutality is coupled with a low cunning in committing it without getting caught by the umpire.[4]

Endicott Peabody, headmaster of Groton Preparatory School and a man whose name does not exactly sing "linebacker," urged the President to convene a meeting of coaches and representatives of Harvard, Yale, and Princeton. Strange as it may seem, these three were not only football powerhouses; they were also teams notorious for rough play, for bending rules, for cheap shots and furtive brutality, for getting away with things contrary to the rules of the game. Whatever claims or pretensions to gentlemanliness these schools exhibited off the gridiron, once their students stepped onto a field between the goal lines the calfskin gloves and politesse were off, replaced by bare knuckles and bloody oaths.

And so on October 9, 1905, TR, "who had never played football," gathered at the White House six representatives of elite football schools — no, not Florida, Texas, and Alabama, but the aforementioned Yale, Princeton, and Harvard.[5] How times change! (Though he never played the game, to borrow the phrase Howard Cosell once applied to himself, TR did much prefer football to baseball, which he considered one step above tiddlywinks.)

The sextet invited to this conclave included Walter Camp and John Owsley of Yale, Bill Reid and Ed Nichols of Harvard, and John Fine and Arthur Hilldebrand of Princeton. Secretary of State Elihu Root was also present, presumably in case war broke out between Harvard and Yale and a mediator was needed. The eight men lunched, then TR stepped out to take care of other matters before asking three of his guests to draft a joint statement pledging their support of fair play and adherence to the rules of the game. Walter Camp would release the draft statement to the press. It read: "At a meeting with the President of the United States it was agreed that we consider an honorable obligation exists to carry out in letter and spirit the rules of the game of football relating to roughness, holding, and foul play, and the active coaches of our universities being present with us pledged themselves to so regard it and to do their utmost to carry out that obligation."[6]

The *New York Times* praised the president in an editorial: "Having ended the war in the Far East, grappled with the railroad rate question...prepared for his tour of the South and settled the attitude of his administration toward

Senator Foraker, President Roosevelt to-day took up another question of vital interest to the American people. He started a campaign for reform in football." There were no areas of human endeavor, it seemed, that were beyond the competency of the president. As sports historian John Sayle Watterson writes, "Far from elevating the tone of football, Roosevelt's intervention only focused public scrutiny on the game — and, for the worse, because it uncorked criticisms that had been bottled up for more than a decade." In fact, "the 1905 season brought about an unprecedented number of recorded fatalities and injuries."[7]

So the 1905 season continued in as violent a fashion as the 1904 season had, despite the president's attempt at brokering football reform. The newspapers carried sensational stories of collegiate brutes punching, kicking, gouging, and otherwise manhandling their foes in a spirit most unlike that of the Marquis of Queensbury. Wild free-for-alls and cheap shots were aplenty; the gridiron was rife with serious injuries. A Harvard center punched a Penn player in the face. A Wesleyan gridder kicked a Columbian in the gut. Most notoriously, a Yale hoodlum, Jack Quill, ignored a Harvard player's (Francis Burr's) fair catch signal and smacked him full on in the face, busting his nose and pouring freshets of blood over the green field. This was too much for TR, whose son Ted had broken his nose and been bruised a week earlier in the Harvard–Yale freshman game. Not only had poor Burr been abused, but Yale had shut out Harvard for the fourth year in a row. This was too much for the head of state to bear! Informed by his Harvard sources that the Elis had played dirty and tried to knock Harvard's top varsity player out of the game, the president put aside petty affairs of state to investigate the Harvard loss.

President Roosevelt summoned Harvard coach Bill Reid to the White House to get to the bottom of the fair-catch incident. The blame was laid upon…the referee, Paul Dashiell, who called no foul on the play. Dashiell was no mere part-time ref; he was chairman of the national rules committee of college football. He was also considered a close ally of Yale's Walter Camp, and thus an enemy of the Harvard gridders. As football historian Watterson writes, "Roosevelt concluded that Dashiell should have called a penalty."[8] Maybe he should have, though Roosevelt, who had never played a down of football, had no standing other than that of a typical Monday Morning Quarterback, and the idea that the President of the United States should express public opinions on the quality of officiating in a college football game is one that would have struck every single one of President Roosevelt's predecessors as nothing less than bizarre. Were there no limits to the power of the presidency, and to the reach of the bully pulpit?

The president wrote a letter to the referee scolding him that his bad call, or no-call, "prevented all chance of Harvard winning" the game.[9] How many fans over the last century would have loved to dress down officials who robbed their teams of victories with letters on official White House stationery?

Although no Presidential Statement on the Rules of Football was issued on White House letterhead, Roosevelt's intervention sent waves through the sport.

As an early historian of American football, Parke H. Davis, put it in 1911, "So acute did the criticism become in the latter season [1905] that Theodore Roosevelt, President of the United States, in the month of October, called to Washington the representatives of football of Harvard, Princeton, and Yale, and impressed upon them the necessity of removing every objectionable feature of play, at the same time giving the sport, if rightly played, the prestige of his endorsement."[10] Now, no president had ever given a sport "the prestige" of the endorsement of the executive of the federal government, for no president to that time had really considered such an endorsement his to give in the character of the President of the United States.

Nevertheless, in the next year Northwestern University and Union College dropped the sport (a player had died from injuries suffered in the Union-Rochester game), as did Stanford and the University of California. The faculty-run Columbia Committee on Student Organizations voted to disband its football team. President Nicholas Murray Butler concurred, and Columbia dropped the sport for the next ten years. (Ivy League fans would question whether Columbia plays football even now, but that's another story.) Harvard president Charles Eliot was also anti-football. Reform, he thought, was impossible — the thing should be junked.

In fact, as Ronald A. Smith writes in "Harvard and Columbia and a Reconsideration of the 1905–1906 Football Crisis," in early December 1905 the abolitionists very nearly won a major victory. Meeting at the Murray Hill Hotel in New York City on December 8, 1905, Columbia was joined by Union, Rochester, Stevens, and New York University in voting aye on the resolution "that the game of football, as played under existing rules shall be abolished."[11] Eight schools voted no — West Point, Wesleyan, Fordham, Syracuse, Swarthmore, Haverford, Lafayette, and Rutgers. So by a vote of eight to five, this set of thirteen influential schools voted to not recommend the abolition of the sport of collegiate football. Had two schools switched sides — let's say Swarthmore and Haverford — the vote would have been aye. Other schools would have been under no obligation to follow their lead, of course, but many probably would have, and the possibility of college football virtually disappearing in 1906 would not have been all that far-fetched.

Strange to think, isn't it? Such football factories as the University of Alabama, Ohio State, and Miami of Florida owe a fair debt to Swarthmore and Haverford.

Still, other schools — New York University, Northwestern — joined Columbia in dropping football. Harvard's dons had wrestled with the appropriateness of football in their institution for decades. President Charles Eliot and the Faculty of Arts and Sciences favored abolishing the sport in 1895, but while the Harvard–Yale games of 1895 and 1896 were canceled, the sport remained. The Faculty of Arts and Sciences voted again to bar the sport in 1905, but by this time reform was in the works. And as Ronald A. Smith writes, a "graduate of Harvard, Theodore Roosevelt, loved football and was in a position to attempt to save the game from possible extinction."[12]

TR's next step was taken just after the December 8 meeting at which collegiate football survived by that eight-to-five vote. For the December 8 meeting led to a December 28 meeting at which the forerunner to the National Collegiate Athletic Association (NCAA) was hatched.

The problem, as of December 1905, was that despite the honeyed words of the October joint statement upon fair play and sportsmanship, the rules of the game still seemed to encourage brutality, at least in the eyes of the reformers. And those rules were devilishly hard to change. For the Intercollegiate Football Rules Committee, dominated by the large Eastern schools, was the roost of secretary Walter Camp, the coach (Yale, Stanford) and tireless publicist, who was and is known as "the Father of American football." Although only in his mid-forties during the 1905 controversy, Camp was the respected eminence grise of football, a man whose integrity was beyond question — except when it was questioned.

Impatient with the obstructionism of Camp's Rules Committee, representatives of 68 football-playing colleges and universities met on December 28, 1905, in New York to form the Intercollegiate Athletic Association of the United States. President Roosevelt, who was in frequent contact with Harvard football coach William Reid — Harvard believed that Camp was intransigently pro-Yale and anti-Harvard — assured the coach that "he would exert his influence in behalf of a joint committee" to reform the rules, writes Guy Lewis.[13] Camp's stranglehold would be broken — with a little help from friends in exceedingly high places.

The President lobbied for a joint session of Camp's Intercollegiate Football Rules Committee and the upstart Intercollegiate Athletic Association of the United States. He pressured poor Dashiell, the referee who had missed the roughing call in the Harvard–Yale game and who also happened to represent the Naval Academy on the rules committee, to support the merger. It would have required an adamantine will and a complete and utter indifference to his professional fate for the representative of the U.S. Naval Academy to defy the president on a matter he cared so deeply about.

Three months of meetings followed. Ronald A. Smith has argued that Harvard's Reid, who was effectively TR's agent in these negotiations, was the key figure in bringing about the desired rule changes. Reid warned the old guard that "the rules go through or there will be no football at Harvard; and if Harvard throws out the game, many other colleges will follow Harvard's lead, and an important blow will be dealt to the game."[14]

Most of the proposed changes were accepted. And so certain basic rules of the game were irrevocably changed: thereafter, it would take ten yards instead of five to make a first down; a neutral zone was established at the line of scrimmage; and, critically, the forward pass came into being.

These seem like logical, even inevitable evolutions in the rules of the game, but that is only in hindsight. There were skeptics about the land in 1906 who wondered if such innovations as the forward pass would in fact kill the game. For instance, Benjamin Ide Wheeler, president of the University of California, declared that the altered rules had created "a practically new game. No man can yet tell what that game will really be. At present it is merely a body of rules on paper. What will be the effect of the rules requiring the side with the ball to make ten yards in three downs, and the rule allowing a forward pass can be established and known only when the proposed game shall have been played for a considerable time." President Wheeler doubted that the game would have that considerable time. "I do not believe the present experiment in American college football can survive," he stated confidently. "In my opinion, the whole country will within five years be playing the Rugby game."[15]

Wheeler was not the first college president to get something disastrously wrong. Nor was he the last. Ten yards became the benchmark. (Four downs instead of three came later.) And the forward pass widened the game in ways perhaps inconceivable to the early reformers. There were, in the beginning, penalties associated with incomplete passes, and it took a while for coaches to figure out how to use them. (Woody Hayes, for one, never did figure it out.) But the fact that presidential intervention — not college presidents, but President Roosevelt — was critical to the achievement of these reforms casts the game, and the history of the interplay between sport and state, in a somewhat different light. If the rules of a boy's game were fair game for government involvement, as Theodore Roosevelt seemed to think, then what else could government do to promote sports in America?

By 1910, the two committees had become one, now operating under the still-extant name of the National Collegiate Athletic Association, or NCAA. By "protecting Harvard's welfare" and using "his position in government and his personal power of persuasion to force the Rules Committee" to cooperate with the reformists, TR deserves to be reckoned among "the founding fathers of the National Collegiate Athletic Association," writes Guy M. Lewis.[16] He is also, indirectly, the father of the forward pass, which resulted from the power shift away from Walter Camp's committee. Whether that was a good or bad change — which may depend upon whether your favorite team has been quarterbacked by Tom Brady or Ryan Leaf — one can legitimately ask whether or not it is a change in which the President of the United States should have any say whatsoever.

"Roosevelt's intervention in organized sports has practically no parallel" in the history of the American presidency, observes sports historian Watterson, though a somewhat less vigorous exerciser, Bill Clinton, intervened unsuccessfully to solve the baseball strike of 1994–1995.[17]

President Roosevelt did not issue any ukases reordering the game of football. He lacked the power. As Watterson writes, "Roosevelt could not have ordered an end to gridiron violence or football even if he had wished to do so. With the exception of the service academies, he had no authority to issue edicts binding the colleges or altering their athletic policies. As Roosevelt clearly indicated, he had no desire to abolish a sport that he regarded as so essential for building toughness and character."[18]

Yet, all the president's meddling triggered a series of conferences involving representatives of football-playing colleges that eventually led, over the next couple of years, to significant, game-altering rule changes intended to open up the game. (Which they certainly did.) Most football fans would say that these changes were for the better, and it is possible that they would have come about had the president never uttered a word about football. But they would not have come about quite so quickly, and the question remains as to whether federal intervention in sport is wise or constitutional policy — even if that intervention leads to Don Hutson, Otto Graham, Jerry Rice, and Peyton Manning.

The Red Team

By the 1920s, sports mania had swept the land. Babe Ruth, Walter Johnson, Jack Dempsey...young and old alike followed the exploits of larger-than-life athletes through the burgeoning media of daily newspapers, radio, and, increasingly, newsreels at the movie theater. Politicians labored to associate themselves with popular athletes and successful teams; baseball, especially, was the all-American game.

Even the communists got into the spirit, though their motives were far from Olympian pure. As Mark Naison explains in "Lefties & Righties: The Communist Party and Sports During the Great Depression," the Communist Party USA was almost entirely a foreign-born movement in the early 1920s. Only one in ten Communists spoke English as a first language. The party organized "networks of sports clubs," but "the sports which they emphasized — soccer, gymnastics, and track and field — were ones which had little attraction for native Americans or even second generation immigrants."[19]

Americans wanted to play and watch baseball and football and boxing. They flocked in the millions to voluntary organizations and leagues, which permitted them to do so. The YMCA, the AAU (American Athletic Union), churches, and ethnic clubs sponsored sports leagues in a nationwide demonstration of the associational voluntarism of which French observer Alexis de Tocqueville had written a century earlier in his classic *Democracy in America.*

Unable to beat Y teams and church leagues, the Communists joined them. They, too, sponsored teams, and for the most part these teams and players fielded fly balls and threw roundhouse rights (or, more likely, lefts) and hiked the ball without the least thought of worker alienation or the dictatorship of the proletariat. The *Daily Worker,* newspaper of the Communist Party USA, even devoted an entire sheet of its eight-page length to sports. Other than the praiseworthy and hardly communistic idea of integrating Major League Baseball, the *Daily Worker* toned down the politics for mostly straight (if, inevitably, red-tinged) coverage of the Yankees and the Monsters of the Midway and the Bronze Bomber.

The communists' infiltration of baseball was not uncontested, of course. Senator Edwin Johnson (D-CO) said in 1953, "If the free world and Iron Curtain countries could compete on the baseball diamond, plans for war would disappear from the face of the earth like an early-morning dew."[20] Set beside the astronomical costs of the Cold War, Senator Johnson's dream of a true World Series could have been the greatest budget savings in the history of the American fisc. But Ivan never got to first base.

"Before 1950, Congress virtually ignored" sports, as John Wilson writes in *Playing by the Rules: Sport, Society, and the State* (1994).[21] Things changed, however, in post-World World II America. The federal government enjoyed unprecedented deference from a citizenry who had recently come through a New Deal and a world war that had drastically changed the relationship between Americans and their central government. A Cold War was incipient, too, and Americans looked to Washington for alms and assistance as never before. It is thus no coincidence that political scientist Arthur T. Johnson, in his study of Congress and American sports, began his tally of sports-related legislation with the year 1951. Between 1951 and 1978, Johnson found, almost three hundred separate congressional bills and resolutions involving professional sports had been introduced. Congress had become, in the words of Stanford economist Roger Noll, "the most likely source of significant change in sports operations," second only to player-management relations.[22]

In the 1950s, much of this legislation involved clarifying the applicability of antitrust law to baseball and other sports. While the antitrust question is beyond the scope of this book, it is instructive, even amusing, to note that the immunity of baseball from antitrust law has been, to mix sporting metaphors, a political football. Members of Congress from cities desiring expansion teams periodically threaten to hold hearings on the subject; the lords of baseball respond by granting franchises to the cities of politically influential figures. It usually works.

When the National Football League (NFL) and its young and exciting competitor, the American Football League (AFL), proposed to merge in 1966, Congress dutifully complied by granting an antitrust exemption — but at a price. Passing the necessary legislation required a certain amount of parliamentary legerdemain to overcome the opposition of Rep. Emmanuel Cellar (D-NY), chairman of the House Judiciary Committee. Two guileful old congressional hands pulled that one off: Senator Russell Long and Rep. Hale Boggs, both Louisiana Democrats. It may or may not have been an incredible coincidence that New Orleans was then awarded an NFL expansion franchise — ironically named the Saints.

A decade later, Senator Warren Magnuson (D-WA) was similarly rewarded for carrying the NFL's water. He committed a noteworthy breach of congressional etiquette by failing to file a conference committee report on an anti-TV blackout bill that had passed both the House and Senate. (The blackout rule permits a team to bar the telecast of its home games to the hometown audience if a certain attendance threshold is not met. The root of the justification for ending blackouts is the ubiquity of public monies in sports.

As Wisconsin Democrat Les Aspin explained, "It would be the height of irony if sports fans were further excluded from viewing their teams while, at the same time, their money was going for the construction and maintenance of bigger and more modern stadiums for their teams to play in.")[23]

In any case, Senator Magnuson sabotaged the blackout bill, leaving "incredulous and angry House conferees" in his wake.[24] But his reward was sweet: the NFL soon welcomed into its exclusive club the Seattle Seahawks. The man who defeated Magnuson in 1980, Republican Slade Gorton, then attorney general of Washington State, had sued the American League when the Seattle Pilots decamped for Milwaukee in 1970. Once in the U.S. Senate, Gorton took up the Magnuson burden of intervening in professional sports team-location decisions. In 1985, Senator Gorton introduced the Professional Sports Team Community Protection Act, which mandated expansion in the NFL and Major League Baseball. Responding to critics who charged that this legislation vastly exceeded the scope of congressional power, Senator Gorton sidestepped the question by pointing out, correctly, that adding more teams to these leagues would lessen the "disparate bargaining power between teams and cities."[25] By awarding teams to such cities as Portland, Las Vegas, Birmingham, and other outliers, which are constantly used to extort new stadiums from taxpayers, the potency of such threats to move would be greatly diminished. Moreover, by accepting antitrust exemption, the leagues had subjected themselves to the Sherman Act's requirement that monopolies not misuse their power. "Surely," declared Senator Gorton, "it is within Congress's authority to find that the prolonged refusal of certain leagues to expand the availability of their product when there are cities which can support a team, is a misuse of monopoly power."[26] Senator Gorton's bill went nowhere; but then, neither did the Seattle Mariners or Seattle Seahawks.

While it is true that granting Congress the power to break down barriers to entry into professional sports leagues would probably improve the bargaining position of local governments vis-à-vis team owners, it would also dilute the quality of play, as political scientist Arthur T. Johnson notes in *Public Administration Review,* and potentially produce a chaotic situation in which teams constantly move in and out of professional leagues, undermining stability and leading to "inferior play." The flipside of the relocation debate — the proposal to restrict team movements — would freeze in place the current setup, which is unfair to cities currently lacking teams. It is also, as Johnson writes, an abridgement of "franchise owners' property rights as well as right of mobility."[27] The legislative cures, it seems, are worse than the disease. This question is revisited at greater length in the fifth chapter.

Despite the examples set by such legislators as Senators Magnuson and Long, tit for tat hasn't always been a successful strategy. Johnson notes that the removal of the baseball Senators from Washington resulted in retaliatory tax legislation from a suburban Virginia congressman, Joel Broyhill — but no expansion team was awarded as appeasement. The 1970s even saw the creation of a House Select Committee on Professional Sports, which many observers speculated "was designed to force baseball back to Washington," though in typically dilatory Washington fashion the return would be delayed for almost three decades.[28]

Even the Great Society set its sights on baseball. Unsuccessful legislation in 1967 would have "made $10 million worth of free tickets available each year for distribution to underprivileged youth."[29] Companion legislation to provide eligible youths with free Sno-Cones, dirt bikes, and pliable dates also failed to pass, but the legislative floodgates were opening.

A Sports Czar?

The most revolutionary idea pertaining to American sports came from Senator Marlow Cook (R-KY), who proposed in 1972 the creation of a Federal Sports Commission (FSC), a bureaucracy whose task would have been to "preserve the system of sports which has provided many years of enjoyment and excitement for millions of American people."[30]

The commission proposal and the surrounding debate suggest just how far down the road of government supervision of professional sports some influential figures were prepared to go. In a strange echo of FDR's Four Freedoms, Senator Cook's bill declared yet another apparently natural right of which Americans had been previously unaware. To wit: "The Congress hereby declares that the public has a right to a stable, financially sound professional sports system, that unstable conditions now exist within professional sports, including the arbitrary sale and transfer of team franchises, the pirating of professional athletes by the various teams and leagues, inequitable arrangements relating to the broadcast of professional sports events on commercial television, inefficient and disruptive mechanisms for bringing amateur athletes into professional sports, and uncertain conditions concerning the forms and provisions of player contracts."[31]

All these manifestations of "instability" — which others might call the mere hustle and bustle of any thriving enterprise — were to be within the bailiwick of the FSC, an appendage of the U.S. Department of Commerce.

This entity was to be governed by three presidentially appointed commissioners, one of whom was to be designated chairman — inevitably to be redubbed "Sports Czar" by the American press. The commission — the FSC, as the acronym went — was to be empowered to promulgate rules under which professional sports would operate. By *rules*, the FSC's backers did not mean that Washington bureaucrats were to determine the degree of contact necessary to trigger a pass interference call, nor was the designated hitter to fall under its capacious umbrella. At least, not at first. Rather, these rules would pertain to franchise moves, player contracts and drafts, and television.

The Senate Commerce Committee held four days of hearings on Senator Cook's Federal Sports Act of 1972, and the record of the proceedings is, at times, illuminating. It is also at times bizarre, thanks in part to such witnesses as Howard Cosell, the egomaniacal sports journalist and color commentator on ABC's *Monday Night Football.*

Senator Cook was not exactly sculpted in the laissez-faire mold of a later Republican U.S. Senator from Kentucky, Rand Paul. Cook declared in introducing his legislation the "right of the sports fan to a stable professional sports system; one upon which he can rely, and one which he can enjoy." His commission and the hearings held to consider it ranged across such concerns as television blackouts, the draft (not the one that conscripted young men into the army and sent them to Vietnam but rather the one that conscripted star athletes and sent them to the Buffalo Bills), franchise relocation, and contracts. It also grazed upon the question of public subsidies for stadium and arena construction, though Senator Cook, as a very modern Republican, scoffed at those throwbacks who held that sports "should remain free of governmental interference."[32]

The related matters of franchise relocations and government-built stadiums recurred throughout the hearings. Philip B. Brown, counsel for the National Collegiate Athletic Association, lit into "the coercive financial leverage" by which owners of professional sports teams "force municipalities into building and financing super sports structures under terms and conditions which are financially unsound," so that "the city and its citizens will be saddled with an annual debt that in no way can be met by rental income or ancillary revenue."

Given that the NCAA was in the process of becoming, in essence, the farm system for the NFL and NBA, this was a spirited protest by Mr. Brown, and it set Senator Cook into a fit of musing about the way that Kansas City/ Oakland Athletics owner Charlie Finley had, in the fairly recent past, flirted with the city fathers of Louisville, where he "got everybody there all excited" at the long-shot prospect of Finley moving his team to Ohio River country. "[W]hen we really wanted to sit down and talk turkey with him," Cook repined, "the turkey was not even on the plate." Instead, as Senator Cook and the Kentucky boosters eventually realized, Charlie Finley "didn't come to Louisville for any idea of considering Louisville as a home for a baseball franchise. He came to have a city make a commitment so he could then take that commitment and force the communities that he really wanted to go to, to increase their bid."[33]

This, as will be seen, is standard operating procedure among professional sports owners, but it only succeeds because city fathers and mothers and the economic development officials who enable them are willing to spend whatever sum of taxpayers' money it takes to lure a franchise, or to keep one whose owner has a roving eye. (In fact, Finley did apply to the American League for permission to move the Kansas City Athletics to Louisville in 1964, but he was voted down by the other American League owners, 9–1. Four years later, those owners approved the team's move to Oakland by a vote of 7–3.)

In Senator Cook's world, almost nothing was to be beyond the scope of the FSC. Among those testifying at the hearings was Angelo Coniglio, a sports fan from Buffalo who was involved with the AFL Identity Committee, an organization of fans with reservations about the AFL-NFL merger. Senator Cook took an inordinate interest in Mr. Coniglio's complaint that "as a season ticket holder for the Buffalo Bills, I was forced to purchase tickets to exhibition games in order to keep my season tickets."[34] Now, this may have been a less than sporting practice on the part of the Bills — as well as the thirteen other teams that also required season ticket holders to buy ducats for meaningless preseason games — but is it really, to use the old phrase, a *federal case*? Ought the ticket-selling philosophy of the Buffalo Bills be a matter of supreme interest to the United States Senate? And if it is, what then is *not* of interest to the Senate?

Mr. Coniglio did bring up, almost as a footnote to the raging controversy over exhibition-game tickets, the not so small matter of Bills' owner Ralph Wilson having threatened to move his woebegone team unless the taxpayers of Erie County built him a new 80,000-seat stadium to replace the long-time home of the Bills, War Memorial Stadium, affectionately known as "The Rockpile," a federally subsidized 1937 project of the Works Progress Administration, or WPA. (New Deal critics called the WPA "We Putter Along.") Coniglio suggested to Senator Cook's committee that perhaps the

Federal Sports Commission could have intervened in Buffalo in some unspecified way — not, presumably, by way of footing the bills for what became, eventually, Ralph Wilson Stadium, built by the County of Erie in 1973 for $22 million, although once the federal government enters a field it seldom leaves its checkbook at the door.

No sports happening of the early 1970s would have been complete without an appearance by Howard Cosell, the mouth that launched a thousand mute buttons, and the "Monday Night Football" announcer did not disappoint. True, he did concede to the committee that "I am not telling you that I possess a monopoly on integrity or brains," but who else, in addressing a question about the NFL, would say that "the football Giants have had an unparalleled prosperity notwithstanding the fact that in nine of the last ten years documentarily they have presented a continuity of gridiron futility"?[35]

Cosell had just flown in from Paris, as he wasted no time in advising the committee. This master of verbosity took his time in getting to his point — "I have lived a contemporary lifetime on the sports beat," he said, emphasizing each syllable as if auditioning for *Plutarch's Lives*, "and I have come to the conclusion that at the very least in principle the kind of bill that you have proposed before the Senate of the United States" — you get the point; he was not short-winded — "is deserving of passage."[36]

Cosell addressed the matter of subsidized stadiums and footloose owners trawling for municipal goodies, though he did so in a grandiloquent manner that perhaps shed more light on himself than on the problem. He accepted implicitly the Chamber of Commerce fallacy that a city is not "major league" unless it hosts major league teams. Of the New York Giants' impending move to the Meadowlands in New Jersey, Cosell bemoaned that it gave "public notice to the notion, transcendental to the importance of sports, that the greatest city in this country is unlivable, which it is not."

In other words, if the football Giants play at a stadium seven miles from midtown Manhattan, New York City is thus marked as unlivable in the eyes of the nation.

How to keep the Giants in New York, and other franchises tethered to the cities with which they have been identified? The "only answer to the problem of carpetbagging in the major professional sports," Cosell opined, was "through the creation of the kind of Commission that you have recommended."[37] Owners who wished to move teams would need to petition the Federal Sports Commission — that is to say, the federal government — for permission, and in its wisdom the FSC would have decided if, say, the St. Louis Cardinals would be allowed to move to Phoenix, or the Baltimore

Colts to Indianapolis. Not that the latter move was even faintly conceivable. Heaven forbid. With the lack of prescience but abundance of certitude for which he was famous, Howard Cosell announced to the U.S. Senate in committee assembled that Baltimore Colts' owner Carroll Rosenbloom, whose ties to gamblers and whose mysterious death would soon make him among the most notorious figures in the history of the National Football League, was "perhaps the finest owner I know in professional sports, certainly the most generous to his players."

Any criticism of Rosenbloom, whose proclivities were well known, Cosell chalked up to "a series of designed vilifications."

Senator Cook and Mr. Cosell — who would later flirt, briefly, with the idea of running for the Senate from New York — then had this colloquy:

SENATOR COOK: You don't think the fans in Baltimore have anything to worry about?

MR. COSELL: They are not going to lose their franchise. There is no way.[38]

It would be difficult to find a more inaccurate prediction in the annals of sports journalism, but then that's the joy of predictions: no one ever calls you on making a bad one. Not even Baltimore Colts' fans.

It is great fun to read the absurdly self-important pontifications of Mr. Cosell, but his remarks also indicate certain trends in the American attitude toward sports and government. The federal government, by the early 1970s, was imagined by many to be capable of performing almost any task, righting almost any wrong. Vietnam was not going well, and Watergate was just over the horizon, and the Great Society had not brought about a great society, but Americans had a faith in the power of Washington that would have startled their ancestors. Cosell actually viewed the movement of teams, and the bidding wars between cities for teams, as an affront to the majesty of the federal government. He said that such activity "results in its own way in a diminution of the dignity of the Senate and the Congress of the United States as politically elected officials in the State of Georgia grabbed for Milwaukee [the baseball Braves] while the politically elected officials of Wisconsin decry that which is going on."[39]

A tug-of-war over the Braves between Atlanta and Milwaukee diminished the integrity of the Senate and the House of Representatives? (The loser, Milwaukee, soon thereafter got another team, the Brewers.) Quite possibly that tug-of-war diminished the city and county coffers in Atlanta and Fulton County, but the integrity of the U.S. Senate and House? Weren't such characters as Senators Harrison Williams (D-NJ) and Vance Hartke (D-IN) and

Representatives Wilbur Mills (D-AR) and Daniel Flood (D-PA) doing that on their own? And this is not even to mention President Richard Nixon and Vice President Spiro Agnew, both of whom within the next election cycle would resign in disgrace.

Not every witness was friendly to the Federal Sports Commission. Commissioner Pete Rozelle of the National Football League — Alvin Peter Rozelle, as Cosell liked to intone — spoke in respectful opposition, as did Major League Baseball Commissioner Bowie Kuhn.

Commissioner Rozelle denied that pro sports are even a big business: "The reality is that professional sports are, in economic terms, very small business indeed."[40] He was right. As Senator Sam Ervin (D-NC) correctly observed, "The professional sports industry is about the size of the pork and beans industry."[41] And yet, Rozelle continued, "professional football is the most oversupervised, overexamined, and overregulated business in America today." As evidence he listed the numerous laws and agencies under which the NFL operated: the Federal Trade Commission, the Federal Communications Act, the National Labor Relations Act, the Price Commission, the Pay Board, the Equal Employment Opportunity Act, and the entire network of employer- and workplace-regulating statutes which define work relations in modern America.[42]

Commissioner Rozelle, when asked by Senator Cook about taxpayer-subsidized stadiums, rehearsed what was already becoming an old familiar and very misleading tune — that "many communities see strong indirect benefits flowing from the stadium." He spoke specifically of Riverfront Stadium in Cincinnati, which had cost the taxpayers $45 million but which, he said, had led to people pouring into the city for games. The "entire atmosphere in Cincinnati," he had been advised by one member of Congress, "had been greatly improved." (Senator Cook agreed, calling the archetypically non-descript cookie-cutter Riverfront Stadium "beautiful," and proving for the millionth time that there is no accounting for taste. There is, however, accounting for stadium expenditures.)[43]

Senator Sam Ervin (D-NC), the wry old constitutionalist who was soon to become an unlikely folk hero during the Senate Watergate Committee hearings, impaled Cook's legislation. Senator Ervin was out of step with the new consensus that everything under the sun and moon was the business of Washington. He took aim at the very premise of the legislation, drawling, "I do not believe the public has a right to a 'financially sound professional sports system' if they don't support teams. I just do not believe that the U.S. Government should be in the business of propping up failing businesses which is the issue that would ultimately confront the Sports Commission in some cases. I opposed the Government's guarantees to Lockheed Aircraft and the subsidies for SST. I believe that in our economic system an ill-managed business or one which people are not willing to pay for should face the possibility of failure, if it can reap the benefit of success."

Senator Sam, far from being a troglodytic embracer of past injustices, scorned the proposed commission as one that would lock in place the status quo and not allow for innovation, for experimentation. What, he asked, was so sacrosanct about "stability"? For "if stability had been paramount there would have been no American Basketball League with its new rules, no American Football League." Moreover, Senator Ervin asked, what if a commission froze into place "the barbarian player-management practices which now exist throughout the sports system"? Ervin decried the lack of free agency in professional sports, asking in what other professions were persons bound to one employer for their working lives? The commission was unlikely to alter such arrangements. "After much reflection," Ervin concluded, "I just don't have much faith that a Federal bureaucracy will protect the rights of the players."[44]

Senator Cook, taken aback by Senator Ervin's lack of faith in the federal bureaucracy, posed what he thought to be a confounding question:

SENATOR COOK: You said if a city doesn't support a team, does it have a right to that team. Let me ask the opposite. If a city absolutely supports a team, does it have to stand a chance of losing that team?

SENATOR ERVIN: As long as we have the free-enterprise system I think it does.[45]

Oops. Turns out not to have been such a trick question after all. On occasion, a city that supports a team may lose that team to another city, perhaps because the municipal government of that other city beggars its taxpayers to provide a new stadium and sweetheart deal for the faithless owner. In most cases — Milwaukee losing the Braves and getting the Brewers, Cleveland losing one version of the Browns and getting another — the bereft city will get a new franchise. But not always. For instance, there is the case of Brooklyn and its Dodgers — but as we shall see, even that case was more complicated than the legend lets on.

Senator Cook was pretty much KO'd by the one-two punch of Senator Sam Ervin and a freshman Republican congressman from Buffalo, Jack Kemp, who played pro football for 13 years and was president and cofounder of the American Football League Players Association.

Kemp, like Ervin, strongly opposed the bill, arguing that the collective bargaining process was a more reliable arbiter of player–owner differences than a federal bureaucracy could ever be. Responding to Senator Cook's assertion that not "all of these problems can be solved at the bargaining table," Kemp said with some exasperation, "what is the great problem that can only be resolved by the U.S. Congress? Just because some problems can be proved to exist, there is no guarantee we are going to find a solution down here in Washington, D.C."

Kemp conceded that, of course, "it is in the interest of the business to have some stability," and that franchise shifts "destroy markets" and "destroy fan appeal.... I don't favor capricious moves, but I have not seen them in professional football."[46]

Although Jack Kemp would later earn a reputation as a most voluble gasbag, in this, his freshman term, he tried to school the unprepossessing Senator Marlow Cook in Econ 101. Hopping aboard his hobbyhorse, Senator Cook asked Rep. Kemp about the requirement of some teams that fans purchase tickets for meaningless exhibition games in order to be able to purchase season tickets.

"I do not particularly like it," replied Kemp. But "I am a consumer in the marketplace, and my buying a product, or my abstention from buying a product happens to determine the success or the lack of success of any product and any service in this country. And, to that extent, I guess my ultimate weapon is to not buy in this case, or to register my disapproval by abstaining from buying those tickets."

This simple answer — asserting the power of the consumer to exert pressure to change ticket-selling practices rather than lodging such power in a federal bureaucracy — left Cook nonplussed. But Kemp was on to weightier matters.

"Are you going to suggest in this bill that the Federal Sports Commission is going to make the decision as to whether or not there should be franchise shifts?" the Buffalo congressman asked Senator Cook incredulously.

Well, um, yes, replied Cook in effect. That is just what the Federal Sports Commission would do, among other things. If an owner got restless and started gazing at the greener pastures of a faraway city, the commission, explained Senator Cook, would "make a determination by examining the records, that a franchise is a substantial investment, that it is giving a good return, and, therefore, cannot be moved."[47] In other words, the federal government would become what President George W. Bush later called "The Decider" of such momentous questions as whether the Houston Oilers would be permitted to move to Nashville. The FSC, as a political entity, would be subjected to fierce political pressure in such an instance from the Senate and congressional delegations of Texas and Tennessee, and it is hard to imagine the commission not ultimately deciding in favor of whichever state had the most influence within the corridors of the federal government.

The decision on franchise movement and placement has to be made somewhere: in this case, the question is whether a politically appointed body would be significantly "fairer" or more just than the other owners and the NFL commissioner, with whom the decision rested in 1972 and still rests. It is worth pointing out, too, that in the case referred to, Houston, while losing the Oilers, was soon thereafter awarded the franchise that became the Texans. The one city of the last three decades that was abandoned by the NFL and never received a compensatory franchise was Los Angeles, the second-largest city in America, and a very special case in that (1) many fans seem to prefer *not* having a team, so that on Sundays they are offered instead the best game of the week on television; and (2) there have been a series of proposals to build stadiums to host NFL teams, but as yet the perfect match of site, funding, and franchise has not been achieved. No one really seems to believe that the NFL has "abandoned" Los Angeles in the way, for instance, that Major League Baseball "abandoned" Brooklyn; if anything, Los Angeles's relative indifference to the NFL has frustrated the league.

When pressed about Buffalo owner Ralph Wilson's threats to move the team to Seattle if the city of Buffalo or the county of Erie didn't build him a new stadium, Kemp hemmed and hawed and said that anyone who "saw the decay of the stadium recognized that Buffalo needed a new stadium." He also offered that Ralph Wilson "is a terrific owner," a sentiment never widely held in Buffalo. So even a politician who praises the free market is, in the end, a politician.[48]

Helping Kemp to pile on Cook were former players Leon Hart, the Heisman Trophy winning end from Notre Dame who played eight seasons for the Detroit Lions and was representing the National Football League Alumni Association, and Bill Dudley, NFL Hall of Fame running back for the Lions, Pittsburgh Steelers, and Washington Redskins and later a member of the Virginia House of Delegates, who was representing the Hall of Fame.

Leon Hart relayed to the committee the results of a survey of NFL alumni on the desirability of the creation of a Federal Sports Commission. The results, he said, were "almost unanimous" in opposition. Among the comments scribbled by players answering the poll were "Government control of sports is one more step in the wrong direction," "don't wish any Government infringement in sports at all," "pro sports do not need

Government controls as it would be like another lump on the camel's back," "sports people should run their own business," and "Government intervention in sports would be most appalling."[49]

It was a rout, rather like sending the Monsters of the Midway onto the gridiron to face eleven spindly Senate aides. Senator Cook, thinking like a politician, asked Leon Hart if the NFL Alumni Association might change its collective mind if the bill also created a pension fund from which all NFL players, past, present, and future, could draw.

Hart was not suckered in: "I think the position is that probably we would still be against the Government getting into and organizing a commission of this nature."

Hart, like Kemp, put his faith in the actions of free individuals:

> [I]t is the general opinion of the membership of the NFL Alumni Association that the public can best be protected if they are allowed to protect themselves. The simplest way is not to patronize that with which they are displeased. Their control over professional sports is whether or not they purchase a ticket. As to the financial stability of pro football franchises, franchises will rely on what the basic supply and demand in that area of influence dictates. No show of Government regulation will provide financial stability. The interests of professional athletes depend, like any other product or service, on what they have to sell.[50]

Well and simply put, this made a stark contrast to the interventionist assumptions underlying Senator Cook's bill. And while that bill would never became law, and an FSC would not join the FCC, FAA, FDIC, and the countless other acronyms in the vast pond of government agencies, certain interventionist assumptions, especially with regard to the necessity of municipalities paying for stadiums and arenas, would soon become the conventional wisdom. In fact, they already *were* the conventional wisdom, for the 1970s were a decade in which public sponsorship of boring cookie-cutter stadiums became as much a part of the passing scene as Walter Cronkite and flared trousers.

Hart and Bill Dudley patiently instructed Senator Cook in the free-enterprise solution to the problems, real and imagined, that inspired his legislation and these rather extraordinary hearings. Senator Cook asked Dudley — "as a former member of the Virginia Legislature" — if "a community ought to be badgered into extreme debt to build a stadium for somebody when it has got a 5-year contract, or a 10-year contract, and all of a sudden, at the end of the period, they walk off and go someplace else?"

"No, I don't believe in badgering anybody to do anything personally, Senator," replied Dudley, encapsulating a philosophy that, if widely adhered to, might prevent all sorts of unpleasantness in this world.

Leon Hart chimed in: "As far as personal opinion on whether or not the club should badger the public to put in a stadium for them and service bonds and so forth, if the public desires not to do so, and there is an economic advantage for the club to move to another town, who has agreed to do so, at a particular economic advantage, then should the club move, it is up to the judgment of someone new coming in, it ought to be, and I agree with Senator Ervin, it ought to be a nice place to start another franchise under the conditions that the previous club left."[51]

And on such a note the hearings more or less drew to a close. The league commissioners and the players had spoken, and they were skeptical that any federal bureaucracy could deliver them from whatever problems afflicted what were, after all, profitable and glamorous ventures. Howard Cosell and Senator Marlow Cook begged to differ, placing their faith in the wisdom and power of the federal government. Senator Cook's Federal Sports Commission would not become law, but it would serve as a striking monument on the roadside from laissez-faire to government promotion of professional sports. Even without an FSC, sports in America somehow endured.

And even some liberals began to worry that the statist pendulum was swinging too far onto the playing field. Senator Howard Metzenbaum, Ohio Democrat and advocate of every intervention under the sun, expressed his unease in a 1986 Senate hearing on antitrust legislation affecting sports: "I feel a bit troubled as I sit here today participating in the first of what may be a series of hearings concerning professional sports. And I note that the Commerce Committee is also spending a great deal of time on this issue. I just left the Budget Committee to come over here. The Nation faces serious problems with the deficit approaching $200 billion. We are in the middle of an arms race, the possible confrontation in the whole nuclear area. Unemployment has gone up, and the U.S. Senate is here debating sports franchises. Why?"[52]

When even Howard Metzenbaum thinks government is overstretching its bounds, that's a sign you're entering strange territory.

Sex, Drugs, and Spin

The intersection of sports and politics takes on the coloration of the ridiculous as well as the sublime. Consider Ari Fleischer, the Washington flack who served as President George W. Bush's press secretary from his inauguration in

January 2001 until July 2003. Having made his name by shading truths and selling the Iraq War, Fleischer left what is euphemistically known as "government service" to trade in on his access as a consultant. Give him some credit, though. Instead of selling out as a lobbyist, he formed Ari Fleischer Sports Communications, which permitted him to indulge a lifelong passion for sports.

Fleischer advises a range of clients, from the NFL to the US Olympic Committee to Penske Racing, on dealing with the press. He sold himself to skeptical jocks by insisting on parallels between the worlds of government and sports, telling them that "the only two institutions in our society that have their events covered live and that have sections of the newspaper dedicated to themselves are America's premier sports leagues and the White House." True enough, though then he reaches: "The president, every speech is covered live. I used to stand on that podium twice a day. NFL head coaches, college coaches, many players — NBA, MLB, NHL, Olympics — intense scrutiny. Coverage for CEOs comes and goes. Coverage for senators, congressmen, comes and goes. Even some governors, it comes and goes. The two institutions are the White House and the premier sports leagues."[53]

The next time a press conference by the coach of the Carolina Panthers is covered live on television, perhaps Fleischer's contention will ring truer. But clients like to be flattered, and what .223 hitter for the Florida Marlins doesn't want to be told that his every pronouncement will be covered with more scrutiny than are remarks by the governor of Florida? (It is quite possible that the .223-hitting Marlin has more to say than your typical Florida chief executive, but that is beside the point.)

Ari Fleischer is living every jock-worshipping geek's dream, and it would be uncharitable to chide him too harshly, but his experience points up the inaptness of sports metaphors when applied to affairs of state, the incompetence of government spin doctors when forced into situations where the coercive powers of government cannot come to the rescue, and the general cluelessness of sports executives who actually desire the imprimatur of political operatives.

Fleischer's first major sports client was Major League Baseball. He was called in after the disastrous House Government Reform Committee's March 2005 investigation into the use of steroids in Major League Baseball. Retired St. Louis Cardinals slugger Mark McGwire, who had shattered the MLB single season home-run record in 1998 by knocking 70 balls out of the park, clammed up in what was widely regarded as among the worst instances of congressional testimony in modern memory. "I'm not going to go into the past or talk about my past," he responded when lobbed a softball by a St. Louis congressman about whether he had played the game with "honesty and integrity."[54]

Leaving aside — for a moment — the question of whether or not the federal government had a proper role to play in this investigation in the first place, Ari Fleischer helped to guide McGwire out of seclusion after his much-ridiculed testimony and back into pro baseball, where he now serves as a hitting coach. The consensus is that McGwire misplayed his return with a series of clumsy answers to the "did you use steroids" question, and yet Fleischer, rather like his old boss Bush, keeps failing upwards.

He also took on as a client the Bowl Championship Series, the alliance which chooses those football-factory universities whose teams will play each year in the top five bowl games. The BCS is despised by most college football fans as a cabal that both prevents a true national championship game and also, at the urging of the richest conferences, excludes in most cases lesser-storied football schools from competing in the top bowls, but Ari Fleischer was used to defending the indefensible. In November 2009, under fire not only from fans but also senators representing such discriminated-against schools as the University of Utah, Fleischer cashed a fat paycheck to flack for the BCS. He failed to come up with any defense beyond pointing to the unpalatability or impracticality of certain alternatives, but hey, he's not a miracle worker.

The Green Bay Packers, too, called upon George W. Bush's mouthpiece when in the preseason of 2008 star quarterback Brett Favre went back and forth deciding whether to play in the forthcoming year. Fleischer's description of the similarities between Favre's indecision and his work at the White House was a classic:

> If you're on the receiving end of a crisis — whether it's an anthrax attack, September 11th, whatever the case may be at the White House — you feel besieged, you feel under tremendous time pressure, and you think through, "How can I explain this and get my point across given all this pressure?" If you're the Green Bay Packers and Brett Favre is announcing he's getting on the plane and coming back to Green Bay, you feel besieged, you feel under pressure, you feel like your whole world is about to come down on you.[55]

Now *there's* a man who keeps things in perspective!

Members of the sporting press were not quite so gullible in buying Mr. Fleischer's line as members of the White House press corps had been back in 2003. ESPN columnist Gene Wojciechowski called him "the guy who orchestrated Mark McGwire's bungled re-entry into baseball... The guy who shills for the galactically stupid and indefensible BCS...The guy who counseled the Green Bay Packers on how to deal with their Brett Favre divorce proceedings."

Fleischer's most recent clients have included Tiger Woods, the golfer whose myriad sexual indiscretions landed him on the front pages of tabloid newspapers and in a sex-addict rehab clinic. Wojciechowski wasn't buying it: "Woods doesn't need to be managed, especially by someone who appears to think the BCS is a brilliant idea. He doesn't require advice from the same person who signed off on McGwire's delusional steroids-didn't-help-me-hit-home-runs explanation."[56] And in fact Woods's "apology" was panned as a robotic and insincere act of fake penance that could have come straight from an image-burnishing computer program. Ari Fleischer was probably surprised by how negatively people reacted to his sports clients. It sure wasn't like this at the White House. But then sports can claim with some degree of credibility that it is a meritocracy, and that men and women succeed on fields of play due to their own hard work and talent. No one — not even the most terminally naïve swallower of civics-class clichés — really believes that government is a meritocracy, or that honesty — which sports fans sometimes demand and sometimes even get — is a prerequisite to success in politics or in political spin-doctoring.

Drugs have been the gateway to government regulation over significant stretches of our economic and personal lives. Sports is no exception. The performance-enhancing drugs controversy of the early twenty-first century permitted entrée to the major leagues by members of Congress, who tended to adopt one of two attitudes when star athletes swaggered before their committees. They were either (1) awestruck nerds basking in the reflected glory of the jocks who, for once, let them sit at their lunch table and listen to their stories of heroics on the gridiron or the ball field; or (2) resentful nerds who, presented with the chance to lord it over the jocks, even intimidate them, did so with an unseemly relish.

It had been more or less common knowledge for years that football and baseball players were using various performance-enhancing drugs. Some of these substances had not been specifically banned by their leagues, though in some cases they were being illegally obtained by the users. The lords of Major League Baseball are widely suspected of having known that there was something fishy about the incredible home run totals of the 1990s achieved by such players as Mark McGwire, Barry Bonds, and Sammy Sosa. Lively balls and poor pitching could not account for a veritable explosion of 60-plus home run seasons, given that in all of baseball's history only twice, in 1927 and 1961, had a hitter knocked 60 baseballs out of the park over the course of a single season.

The aforementioned McGwire shattered the MLB home run record by hitting an astounding 70 HRs in 1998. The league played up the feat for all it was worth. The fact that Androstenedione, or Andro, a muscle-enhancing product that had been banned by the International Olympics Committee and the NFL (but not by baseball), had been found in McGwire's locker did not much detract from the celebration. After all, Babe Ruth drank like a fish and ate like a rhino. Each era has its vices. And steroids had undeniable benefits, especially in the way they help athletes recover from the injuries and bruises that are part of the game. But by the time Barry Bonds hit 73 home runs in 2001, the writing was on the pharmacy mirror. A federal investigation into the Bay Area Laboratory Cooperative, or BALCO, linked Bonds with THG, which was BALCO's signature steroid. The press, which had looked the other way when the home run derbies had been good copy, took to sanctimoniously decrying the "cheaters" who gained a chemical advantage by using steroids. And once sanctimonious decrying begins, politicians cannot be far behind.

President George W. Bush, a former co-owner of the Texas Rangers and employer of Sammy Sosa, Rafael Palmeiro, and Juan Gonzalez, each of whom has been, to put it mildly, an object of interest in the matter of performance-enhancing drugs, took to his pulpit to bully baseball. In the 2004 State of the Union address, given at a time when his country was waging a war which he had launched under murky circumstances, Bush stated:

> To help children make right choices, they need good examples. Athletics play such an important role in our society, but unfortunately, some in professional sports are not setting much of an example. The use of performance-enhancing drugs like steroids in baseball, football, and other sports is dangerous, and it sends the wrong message, that there are shortcuts to accomplishment and that performance is more important than character. So tonight I call on team owners, union representatives, coaches, and players to take the lead, to send the right signal, to get tough, and to get rid of steroids now.[57]

With that, the deluge was on. The congressional hopper was filled with bills demanding federal regulation of drug-testing in professional sports. This would have marked a major reversal in labor policy. As Brent D. Showalter wrote in a 2007 paper on steroid testing in the *Marquette Sports Law Review*, "drug-testing of employees is a mandatory subject of collective bargaining" according to the mandate of the National Labor Relations Board — itself a federal entity, so we are hardly talking about a laissez-faire system here. Prior to 2002, the Major League Baseball Players Association had resisted testing for steroids on grounds of privacy, calling such testing "an abuse of human rights."[58] But human rights and privacy don't matter much to members of Congress when TV cameras are present, so by 2005 the rush to regulate was on.

Bills with mom and apple pie titles proliferated: the Drug Free Sports Act, the Clean Sports Act of 2005, the Professional Sports Integrity Act of 2005, the Professional Sports Integrity and Accountability Act. These bills mandated federal drug-testing for major league sports players, though they differed on such relatively minor matters as frequency of testing, punishments for those testing positive, and bureaucratic responsibility for administration. Some limited their scope to the agreed-upon "major" sports, while others extended the federal drug-testing dragnet over the Arena Football League, the Women's NBA, major league soccer, and even minor league baseball. All of them were "considerably more stringent on steroid testing in professional sports than even the harshest policy formed by any individual league."

Senator Jim Bunning (R-KY), the Hall of Fame Philadelphia Phillies pitcher and legendary grouch, sponsored the Integrity in Professional Sports Act. He explained that because, in his view, union and management would not come up with a drug policy that is "satisfactory" to outsiders, especially headline-seeking congressmen, "we're going to move ahead in Congress."[59] Whether or not a player rubbed a steroid balm into his skin was a matter far too critical to be left to collective bargaining. It required the application of the wisdom of 535 solons on Capitol Hill.

Baseball got the message. Knuckle under or let Congress in the front door. So by spring 2006, union and management announced the Major League Baseball Joint Drug Prevention and Treatment Program. Adopted, essentially, under the gun of federal threats and intense media pressure, the policy banned dozens of steroids and stimulants. All players were subject to random tests during the season. Punishment for testing positive for any of the various substances is harsh: a player found to have used steroids is suspended for 50 games after his first positive test result, 100 games after his second, and in the event the dullard does not get the message — or can't hit without being juiced — a third positive test earns him a lifetime ban from Major League Baseball, where he can join Shoeless Joe Jackson and Pete Rose in the Coventry of the immortals.

Had this policy been implemented by the federal government, it might be subject to constitutional challenges on Fourth Amendment (unreasonable search and seizure) grounds, but as an agreement reached "voluntarily" by two private parties it belongs to that class of acts which, although primarily the product of government threats, are beyond the protective reach of the Bill of Rights.

Nor does the Bill of Rights protect taxpayers from the rip-offs and threats and depredations of team owners and the politicians who sedulously appease them. And having demonstrated that government has long blundered into the field of sports, that is the expensive, even extortionate, matter to which we now turn.

Notes

1. John Sayle Watterson, *The Games Presidents Play: Sports and the Presidency* (Baltimore: Johns Hopkins University Press, 2006), p. 37.
2. Guy M. Lewis, "Theodore Roosevelt's role in the 1905 football controversy," *Research Quarterly*, Vol. 40, No. 4 (December 1969): 719.
3. Ibid.: 719.
4. Theodore Roosevelt, "Functions of a Great University," June 28, 1905, *A Compilation of the Messages and Speeches of Theodore Roosevelt*, Vol. 1, ed. by Alfred Henry Lewis (Bureau of National Literature and Art, 1906), p. 644.
5. Watterson, *The Games Presidents Play: Sports and the Presidency*, p. 54.
6. Guy M. Lewis, "Theodore Roosevelt's role in the 1905 football controversy": 720.
7. Watterson, *The Games Presidents Play: Sports and the Presidency*, p. 56.
8. John Sayle Watterson, *College Football: History, Spectacle, Controversy* (Baltimore: Johns Hopkins University Press, 2000), pp. 71–72.
9. Guy M. Lewis, "Theodore Roosevelt's role in the 1905 football controversy": 722.
10. Parke H. Davis, *Football: The American Intercollegiate Game* (New York: Scribner's, 1911), pp. 110–11.
11. Ronald A. Smith, "Harvard and Columbia and a Reconsideration of the 1905–06 Football Crisis," *Journal of Sport History*, Vol. 8, No. 3 (Winter 1981): 8.
12. Ibid.: 10.
13. Guy M. Lewis, "Theodore Roosevelt's role in the 1905 football controversy": 722.
14. Ronald A. Smith, "Harvard and Columbia and a Reconsideration of the 1905–06 Football Crisis": 15.
15. "Football is Doomed, Says Mr. Wheeler," *New York Times*, September 23, 1906.
16. Guy M. Lewis, "Theodore Roosevelt's role in the 1905 football controversy": 724.
17. Watterson, *The Games Presidents Play: Sports and the Presidency*, p. 57.
18. Watterson, *College Football: History, Spectacle, Controversy*, p. 66.
19. Mark Naison, "Lefties and Righties: The Communist Party and Sports During the Great Depression," *Radical America*, Vol. 13, No. 4 (1979): 49.

20. Arthur T. Johnson, "Congress and Professional Sports: 1951–1978," *Annals of the American Academy of Political and Social Science*, No. 445 (September 1979): 110.
21. John Wilson, *Playing by the Rules: Sport, Society, and the State* (Detroit: Wayne State University Press, 1994), p. 24.
22. Arthur T. Johnson, "Congress and Professional Sports: 1951–1978": 103.
23. Ibid.: 113–14.
24. Ibid.: 112.
25. Slade Gorton, "Professional Sports Franchise Relocation: Introductory Views from the Hill," *Seton Hall Legislative Journal*, Vol. 9, No. 1 (1985): 4.
26. Ibid.: 5.
27. Arthur T. Johnson, "Municipal Administration and the Sports Franchise Relocation Issue," *Public Administration Review* (November/December 1983): 525.
28. Arthur T. Johnson, "Congress and Professional Sports: 1951–1978": 109.
29. Ibid.: 110.
30. Ibid.: 111.
31. "Federal Sports Act of 1972," Hearings Before the Committee on Commerce, United States Senate (Washington, D.C.: U.S. Government Printing Office, 1973), pp. 3–4.
32. Ibid., pp. 1–2.
33. Ibid., pp. 44–45.
34. Ibid., p. 54.
35. Ibid., pp. 101, 97.
36. Ibid., p. 94.
37. Ibid., pp. 98, 96.
38. Ibid., pp. 102–103.
39. Ibid., p. 96.
40. Ibid., p. 135.
41. Charles C. Euchner, *Playing the Field: Why Sports Teams Move and Cities Fight to Keep Them* (Baltimore: Johns Hopkins University Press, 1993), p. 65.
42. "Federal Sports Act of 1972," Hearings Before the Committee on Commerce, United States Senate, p. 136.
43. Ibid., p. 155.
44. Ibid., pp. 138–40.
45. Ibid., p. 146.
46. Ibid., pp. 204, 206.
47. Ibid., pp. 209, 205.
48. Ibid., p. 200. Ralph Wilson, an original AFL owner, expresses the old school view of football: "I got into this because I liked football, not because I wanted to make money. I think a lot of owners used to feel the same way. But today, the thing has swung around. We go to league meetings, and we don't discuss football as much as we discuss business: franchise relocations, revenue sharing, sweetheart leases. I don't like it. I don't think it is good for the game." Yet Wilson is old school — that is, circa 1970 — in another way: Ralph Wilson Stadium was paid for and built entirely by the taxpayers, and not by Ralph Wilson. Quoted in Robert A. Baade, "Evaluating Subsidies for Professional Sports in the United States and Europe: A Public-Sector Primer," *Oxford Review of Economic Policy*, Vol. 19, No. 4 (2003): 588.

49. "Federal Sports Act of 1972," Hearings Before the Committee on Commerce, United States Senate, pp. 248–49.
50. Ibid., pp. 253–54.
51. Ibid., p. 260.
52. John Wilson, *Playing by the Rules: Sport, Society, and the State,* p. 27.
53. Zach Berman, "Ari Fleischer, From the Briefing Room to the Locker Room," *Washington Post,* July 22, 2009.
54. "McGwire mum on steroids in hearing," cnn.com, March 17, 2005.
55. Zach Berman, "Ari Fleischer, From the Briefing Room to the Locker Room," *Washington Post.*
56. Gene Wojciechowski, "Fleischer PR won't rebuild Tiger's image," www.espn.com, March 11, 2010.
57. George W. Bush, "Address Before a Joint Session of the Congress on the State of the Union," Jan 20, 2004, www.presidency.ucsb.edu.
58. Brent D. Showalter, "Steroid Testing Polices in Professional Sports: Regulated by Congress or the Responsibility of the Leagues?" *Marquette Sports Law Review,* Vol. 17, No. 2 (Spring 2007): 655, 658.
59. Ibid.: 663–64.

Chapter 3

Parks and Stadiums Until 1960

The American Game, Born Free: Sort of

Before they were ballparks, they were "baseball grounds," and since people seem naturally to be drawn to watching sports played at a high level of skill, those who sponsored the game soon got the idea of fencing off the grounds "for the sole purpose of keeping freeloaders from viewing the game," as Michael Benson writes in his encyclopedic *Ballparks of North America* (1989).[1]

Harold Seymour, scholar and former batboy of the Brooklyn Dodgers, in his standard history of *Baseball: The Early Years*, describes one of the first entrepreneurial ventures in stadium construction. In 1862 — coincidentally, the same year that, as legend has it, President Abraham Lincoln and his son Tad attended a baseball game in Washington, DC — William H. Cammeyer of Brooklyn drained a pond which he had been using for ice skating. He filled it with dirt, graded it, laid out a baseball diamond, and enclosed it. Although the ballpark, called Union Grounds, was "primitive" by later standards, contemporaries regarded it as a fine adornment to the City of Homes and Churches, as Brooklyn was then known. Cammeyer, father of the enclosed baseball park, put in benches enough to hold 1,500 spectators and built on the six-acre grounds a clubhouse and a saloon. A seven-foot-high fence girdled the park, to which Cammeyer charged admission. If you wanted to watch the game, you had to pay a dime. At the opening game in Cammeyer's park, on May 15, 1862, as the nation was riven by a Civil War, a band boomed out "The Star-Spangled Banner," inaugurating a tradition that carries to this day.[2]

J.T. Bennett, *They Play, You Pay: Why Taxpayers Build Ballparks, Stadiums, and Arenas for Billionaire Owners and Millionaire Players*, DOI 10.1007/978-1-4614-3332-3_3,

Fencing in ballparks and charging admission — the American "enclosure movement" — spread rapidly throughout a baseball-mad country. Baseball, at least when the game was played at its highest level, was becoming a business. The amateurism of the prewar era was giving way to semi-professionalism and then professionalism; teams which had once selected players at least partly on the basis of their clubbability and social standing were now looking for the very best players they could find, social background be damned.

The first truly all-professional team, the Cincinnati Red Stockings, went pro in 1869. Largely imported from the East, this collection of "skilled artisans and clerks" whose off-field trades included bookkeeper, hatter, marble cutter, piano maker, and jeweler, was assembled by an English-born cricket master named Harry Wright, who was affiliated with the Union Cricket Club of Cincinnati.[3] Wright organized the Red Stockings shortly after the Civil War. In 1868, his baseball club put fences up around the Union Cricket Club's grounds and charged admission to see the superb ballclub he had assembled. The team played home games at the Union Cricket Club, where nonmembers were charged ten cents per head admission, and undertook a national tour en route to a final record of 57 victories, one tie, and no losses. Even George Steinbrenner at his most mercurial would have been satisfied with that season.

The Red Stockings' theme song began: "We are a band of baseball players from Cincinnati city/ We come to toss the ball around and sing to you our ditty..."[4]

It is hard to imagine any self-respecting ballplayer trilling such a tune, and in any event, only one member of the ten-man Red Stockings was a Cincinnatian: first baseman Charles Gould. The others were imports, mostly from New York City.

The game was evolving, largely through the voluntary and cooperative efforts of thousands of ballplayers from Manhattan to California. They created leagues, organized teams, and played games in stadiums, in pastures, and in parks. Government had very little to do with the evolution of baseball, although the Civil War did deprive teams of players and temporarily cool baseball fever. But as the war ended, the fever broke out again, and no section of the country was immune. Players, "taking advantage of a good market and seeking out the highest bids for their services in a classic laissez-faire fashion," moved between teams, looking for the best deal.[5] This brief period of laissez-faire was curtailed when in 1879 the new National League introduced the "reserve clause," which would, as amended over the years, effectively bind players to a single team, until that team released or traded them. The reserve clause would prohibit player free agency until 1975.

"The formative years of sport after the Civil War saw a freewheeling mixture of entrepreneurialism and voluntarism," as John Wilson wrote.[6] Now and again in baseball's early years, however, government did extend its very visible hand into the workings of the baseball market. The results were seldom edifying. For instance, Boss Tweed, the infamous political spoilsman of New York, owned the New York Mutuals, members of the National Association of Base Ball Players, the leading organization of baseball teams in the 1860s. "Players on the Mutuals were all well paid," writes Harold Seymour, "because Tweed saw to it that they were put on the city payroll" to the aggregate tune of $30,000 per year.[7] Although these were for the most part no-show jobs, they were hardly glamorous: most were in the city morgue, while others manned — or unmanned — the sanitation department.[8] By 1871, William Cammeyer, entrepreneur of Union Grounds, had taken over the Mutuals from the imprisoned Tweed.

Most club owners were an ethical cut above Boss Tweed, and even those who weren't particularly rectitudinous lacked his direct access to spoils and graft. They had to pay the players and build the parks themselves. The owners of the Buffalo Bisons built a 5,000 seat stadium in time for the 1884 season for a grand total of $6,000; the St. Louis Browns built 10,000-seat Sportsman's Park in 1893.[9] New wooden parks went up in Chicago, Brooklyn, Cleveland, Philadelphia, and many other cities of the National League, the rival American Association, and the other leagues that arose in a baseball-crazy country. Ticket prices were generally 50 cents for a National League game and 25 cents for an American Association contest. Owners employed various ways to pack their parks and make a profit: promotions that ranged from clever to ridiculous, bands, Wild West shows, plenteous booze, and even forcing players to serve double-duty as ticket-takers at the gate. (Imagine the double-take if, as one enters Yankee Stadium, Derek Jeter rips one's ducat — now *that* would be worth the inflated price of admission.) But they did not beg cities to beggar taxpayers to build these stadiums for them. (Although in 1870 the Chicago White Stockings of the National Association of Professional Baseball Players played their games on city-owned lakefront property.)

Professional baseball has never been simon-pure, or wholly aloof from grimy politics. While not bold enough to demand public funding, the owners and operators of the "golden age" pulled wires when they could. As baseball historian Steven A. Riess notes in "The Baseball Magnates in the Progressive Era: 1895–1920," during the first two decades of the twentieth century, "seventeen of the eighteen American and National League baseball teams were run by people with significant political connections. These club owners included political bosses, friends and relatives of men in what we could call high political places, and political allies like traction magnates and professional gamblers."[10] Although baseball is beloved in the American memory as a pastoral sport played in meadows by rustic youths, even the early professional teams were the property of urban politicians, often boasting — or hiding — dubious connections.

These well-connected insiders weren't able to exploit their connections for free stadiums, but they could obtain favorable real-estate assessments, tips about available land, breaks on "amusement" licensing fees, confidential information of the sort known first to government functionaries, and advantageous routing of trolley or subway or bus lines. "Transit companies were important supporters of professional baseball in all parts of the country," writes Riess, for the streetcar was an effective and profitable method of getting fans to the park.[11]

The very idea of nonlocal baseball leagues would be unthinkable without innovations in transportation: railroads, then airplanes, to carry teams from one city to another; and trolleys, subways, trains, and automobiles to ferry fans from their residences to the stadium.

In Atlanta, for instance, the first park to host a professional team, Peters Park, was built in 1885 by the Atlanta Street Railway. Consolidated Railway, a competitor, built the Athletic Grounds in Atlanta nine years later in what Michael Benson calls an act of iron horse "one-upmanship."[12] The parks housed the Atlanta Crackers of the Southern League, a team evidently unafraid of un-p.c. monikers.

As sports historian Steven A. Riess writes in "Historical Perspectives on Sports and Public Policy," although the early owners of professional baseball teams "did not employ their influence to get the municipality to build them ballparks. . .they did use their clout to get cities to subsidize their investment in more modest ways."[13] This took not only the aforementioned forms of police and security protection, transportation decisions, inside information about real-estate transactions and opportunities, and the granting of beer licenses and permits, but also the legalization of Sunday baseball.

Riess has studied the decades-long struggle to legalize Sunday baseball. It makes for a revealing, if sometimes quaint, case study in the interplay of government, baseball, and public opinion.

As Riess writes in *The Maryland Historian,* prior to 1890 professional baseball games "were seldom played on Sunday due to puritanical blue laws."[14] The dominant Protestant religious traditions in America held Sunday to be a day of rest — enforced rest, if need be. But immigrants of Catholic, Lutheran, and Jewish faiths desired to recreate, not repose, on Sundays, and their increasing political clout in the 1890s set the stage for a contest whose outcome could be predicted by anyone with a handy copy of the U.S. Census.

In 1880, the National League had actually ejected Cincinnati (with its large German Lutheran population) for violating the league's no-Sunday-baseball policy. The rival American Association, seeking to lure fans from the senior circuit, permitted Sunday baseball in the 1880s. The fin de siècle saw teams cautiously testing the waters. In Chicago, as Riess notes, pro baseball on Sunday was more or less smuggled in during the 1893 season while the sabbatarians concentrated their censorious fire on the decision of Chicago authorities to permit the World's Fair, which Chicago hosted in that year, to open on Sundays.

Opposition to Sunday baseball in Chicago was organized by the International Sunday Observance League (ISOL), which would have found the whole *Bull Durham* notion of a "church of baseball" blasphemous. Church was church, and baseball was baseball, and never the twain shall meet. The ISOL argued in the courts that Sunday baseball games "were a public nuisance that had caused a sharp decline in property values and encouraged the establishment of several disorderly saloons in the vicinity of the ball park." (Interestingly, the establishment of orderly saloons in the vicinity of a ball park is today put forth as a reason for state subsidization.) The league (of churchgoers, not ball throwers) succeeded in having members of the Chicago team arrested on June 23, 1895, for playing on a Sunday. The players finished the game that day after posting bond. Three months later, they were found guilty, for which each Sabbath-breaker was fined four dollars, but the ruling was later overturned, and never again would sabbatarians scotch Sunday baseball in Chicago.[15]

In New York City, too, immigrants and their mostly Democratic political representatives pushed for the legalization of Sunday baseball. Tammany Hall was tightly bound with professional baseball, but Upstate Republicans frowned upon desecrations of the Sabbath. By the early 1900s, New York City teams were evading laws against professional baseball on the Sabbath by charging for scorecards but not admission — a technical point upon which sympathetic local judges and police officers could fasten. The use of "donation boxes" and games played to "support the troops" in the First World War, too, provided an escape of sorts from these prohibitory laws, but widescale Sunday professional baseball had to wait upon action from the state legislature. Finally, after the war to end all wars had ended, and the troops came home, hungry for baseball, the New York State legislature enacted a local option bill that threw open the Sabbath to balls and strikes and the metaphorical killing of umpires.[16]

The last major league cities to permit Sunday baseball, writes Riess, were Boston (1929), with its Puritan tradition, and the Pennsylvania combination of Philadelphia and Pittsburgh in 1934. The South, which lacked major league teams, was resistant to Sunday baseball until the 1930s, with such obvious exceptions as New Orleans, where the good times, not to mention the bunts, are always rolling, and had been on the fields of green since 1880.

Burning for Baseball

By the early twentieth century, owners in the American and National Leagues were replacing the wooden grandstands, which ranged from the rickety to the magnificent, with concrete and steel. Wood, among its other properties, burns. And so the owners learned basic lessons in stadium construction the hard way, when in 1894 a full third of major league parks caught fire at some point.[17] In all, 15 new stadiums went up between 1909 and 1923, including several of the greatest and most beloved parks in the game's history.

The first of the concrete and steel fireproof stadiums was Shibe Park in Philadelphia, home of the Athletics beginning on April 12, 1909. Shibe was a half-million dollar ballpark that seated over 20,000. Team owners Ben Shibe and Connie Mack — the manager who was famous for, among many other things, being the last to manage his team in civilian duds rather than a team uniform — built Shibe Park without going to the city government to beg for construction monies. They started a trend, for as historian of baseball parks Michael Benson writes in his definitive *Ballparks of North America* (1989), "Shibe Park was so big and beautiful that every big league owner wanted one just like it... The era of the wooden ballpark was over."[18]

The Athletics would play in Shibe (later renamed Connie Mack Stadium — it helps to outlive a stadium's namesake) until 1954, when that sadsack franchise moved to Kansas City, but the City of Brotherly Love's National League team, the Phillies, called Shibe home from 1938 to 1970, when they moved into the characterless $50 million Veterans Stadium, which Benson rates "your basic cookie-cutter, with blue–green artificial turf, and yellow and orange seats — like playing baseball in a big McDonald's."[19] Unlike Shibe, Veterans Stadium was paid for by the taxpayers of Philadelphia, who approved a pair of bond issues for the overbudget but underwhelming field of nightmares. The turf was compared to concrete or worse; many seats were remote from the playing field; and the concrete monstrosity was as charmless as a week-old cheesesteak. When, finally, Philadelphia demolished Veterans Stadium on March 21, 2004, few if any tears were shed. The $50 million monument to bad architectural taste and the bilking of taxpayers was razed in just 62 seconds — a record, it was claimed. The late and unlamented Vet was succeeded by Citizens Bank Park, a more attractive venue than Veterans that was built with a public–private partnership for $346 million, split more or less down the middle. The public portion was covered by a car rental tax, while the Phillies, who are owned by a group led by Bill Giles, supplied the private half.

At the other end of the Keystone State, owner Barney Dreyfuss built Forbes Field for his Pittsburgh Pirates. Forbes was the luxury stadium of its day, complete with brass nameplates on box seats and maids in the ladies room.[20] Dreyfuss bought a section of land that contained cows, a livery stable, and a hothouse, according to Michael Benson. This was not exactly a stab in the dark: his friend and advisor Andrew Carnegie offered his two cents, which was worth considerably more than two cents. The *Reach Guide* of 1910 spared no superlative in its assessment of Forbes Field, which opened for business in 1909: "Words must also fail to picture to the mind's eye adequately the splendors of the magnificent pile President Dreyfuss erected as a tribute to the national game, a beneficence to Pittsburgh and an enduring monument to himself. For architectural beauty, imposing size, solid construction and for public comfort and convenience it has not its superior in the world."[21] In 1933, owner Art Rooney's Pittsburgh Steelers of the National Football League would move into Forbes as well. In time, as discussed in the next chapter, it would give way to Three Rivers Stadium, the charmless government project.

Among the quartet of burning ballparks in 1894 was Union Park in Baltimore, home of the Baltimore Orioles of the National League. In classic early baseball fashion, Union Park was built on land leased by Harry Von der Horst, the Baltimore brewery heir (Eagle Brewery and Malt Works) who owned the team. Von der Horst was a good businessman who thoughtfully included picnic tables within the park for hungry fans and on special days supplied each fan with a picnic lunch and a schooner of Eagle beer. Would that such owners still existed!

"He charged a nickel to get in and made $30,000 in 1893 alone," writes Benson, and his talent for supplying what fans wanted even led to a league rule change. The beer garden he set up in Union Park proved popular with players, too, leading to the rule that ballplayers "must come in from the field and seat themselves on the players bench at the conclusion of their half in the field."[22] Had the rule not been in place by the time Babe Ruth hit his stride, there might not have been enough beer to keep Yankee Stadium supplied.

After Union Park burned in 1894, Harry Von der Horst rebuilt it. He didn't whine and moan to the city of Baltimore, nor did he ask the state legislature of Maryland for an appropriation. He and the beer drinkers of Baltimore built it, and led by such legends as Wee Willie Keeler (who famously said "hit 'em where they ain't"), the Orioles were the toast of Baltimore.[23]

Polo Grounds and Dandy Yankees

In Manhattan, New York Giants' owner John T. Brush leased land from Mrs. Harriet Coogan on which to build the final iteration of the famed Polo Grounds on Coogan's Bluff. The Giants had played on these grounds, which had been in the Coogan family for generations, since 1891. The team's wooden grandstand was demolished by one of the spate of early baseball fires in April 1911 — the blaze broke out in the morning, so no one was hurt — and so Brush, an Indiana clothier, determined to construct a "mammoth structure" whose plans were so elaborate as to move the *New York Times* to forecast "the greatest amphitheater and stadium in the history of the city."[24] The $100,000 ballpark was fireproof, of course, made of steel and concrete.

Brush was not among the better-liked owners in professional baseball; one somewhat purple-penned critic said of him that "Chicanery is the ozone which keeps his old frame from snapping, and dark-lantern methods the food which vitalizes his bodily tissues."[25] A man not without ego, Brush attempted to give the ballpark the euphonious — or at least deserved, given that he was footing the bill — name of Brush Stadium, but old habits die hard, and since three previous parks going under the name of "Polo Grounds" had stood in this vicinity, the fans and press insisted on calling the new stadium the Polo Grounds.

The name stuck. Incredibly, this fourth and final Polo Grounds opened for business just two and a half months after the devastating fire.

The New York Giants are said to have handed out free passes to clergy in order to raise the general level of the sport's audience. Whether that worked is an open question. But when John T. Brush's son-in-law sold the Giants for $1 million in 1919, it was to stock broker Charles Stoneham, a Tammany Hall man who later beat indictments for mail fraud and perjury, though his reputation was never exactly fragrant. Stoneham's son, Horace, would later move the Giants three thousand miles from the Polo Grounds to the city by the bay, San Francisco.

The park would also serve as the home of the New York football Giants from 1925 to 1955, and the New York Jets of the American Football League from 1960 to 1963, among other teams. But its most famous tenant was the most storied — and loathed — franchise in all of sports.

The Polo Grounds hosted the New York Yankees from 1913 to 1922, before the House that Ruth Built rose in the Bronx. The Yankees' first home was Hilltop Park, which was not obtained without a struggle. Tammy Hall politico and shady realtor Andrew Freedman — called "the most hated team owner in baseball history" by BaseballLibrary.com (and that's saying something!), and the owner of the New York Giants until he sold out to John Brush in 1902 — had bought up and hoarded the leases to the likeliest spots upon which any competing ballclub might site a stadium.[26] Thanks to Freedman, the Polo Grounds was the only game in town. "In New York," writes Benson, "without the proper connections, it could be next to impossible to secure a suitable site for a ballpark."[27]

In New York City, especially, baseball was never more than a rifle-armed right-fielder's peg home away from politics. The Democratic machine of Tammany Hall hummed with baseball activity; if the corpulent bosses couldn't actually play the sport, they could use their privileged political positions to queer real-estate deals and redirect transit lines and salt opposition in local legislative bodies to interlopers who might want to put a team in what Tammany regarded as its domain. The right combination of land and mass transit could be the key to a profitable ball team, and the savviest inside political players always knew where those passkeys were located.

Andrew Freedman had purchased a majority interest in the team for $48,000 in 1895, or two years before the Tammany Hall moneyman lent his considerable talents to the national Democratic Party as its treasurer. As Steven A. Riess says, Freedman "ran his baseball team as if it were an adjunct of Tammany," bullying players, other owners, and — especially — potential competitors within the City of New York.

Not only did Freedman have most of the likeliest locales for a ballpark locked up, but he also had a remarkably crude willingness to throw elbows at potential competitors. Even in the improbable event that a rival owner might find a decent site for a field, writes Riess, "invaders knew that Freedman would use his political clout to get streets cut through their property or disrupt their transit facilities."[28]

The new American League franchise in New York, which had been relocated from Baltimore, finally located a promising site at 142nd Street and Lenox which Freedman had not yet locked up. The league put together a rather complicated financing proposal that involved the backing of the Interborough Rapid Transit (IRT) Company, the private firm that operated the then-young New York City subway system. The man who had funded the construction of the subway, August Belmont Jr., financier son of the financier and Democratic politico August Belmont, agreed to help, but the plan was derailed by the opposition of a director of the IRT — none other than Andrew Freedman. Already, baseball parks were subject to political manipulation, and with no team was this truer than the New York Highlanders, who would soon become the New York Yankees. (Just for laughs, perhaps, or maybe just in keeping with his generally unpleasant and truculent personality, Freedman also used his position with the IRT to block a line that would have served the Brooklyn Dodgers. Brooklyn was not Manhattan, and there was not a great overlap in their fan bases, but Freedman seemed not to want to waste an opportunity to do even a fairly distant competitor an ill turn.)

Foiled by Freedman and Tammany Hall in the search for a suitable home for its New York team, American League president Ban Johnson "realized he needed a political bigwig (i.e., gangster) on his side to do battle with Freedman," writes Michael Benson.[29] On the twin theories that no major sports league could succeed without a team in New York City and that if you can't beat 'em, sell out to 'em, Ban Johnson awarded New York's American League franchise for $18,000 to a nattily dressed Tammany Hall figurehead and coal dealer named Joseph Gordon. Gordon was fronting for two dubious characters: a gambler named Frank Farrell and a former New York City police chief named William Devery, whose "regime was so blatantly corrupt that in 1901 the state legislature decided to abolish the position and replace it with a commission system."[30]

The team leased property belonging to the New York State Institute for the Blind (which is not where the American League trained its umpires, despite later claims) in Washington Heights between 165th and 168th Streets. The 10-year lease was pegged at $10,000 per year.

Thomas F. McAvoy, a Tammany stalwart, received contracts to excavate the site ($200,000) and build, for $75,000, the much-derided Hilltop Park, a bare-bones, somewhat cheaply made 16,000-seat stadium, which would host the New York Highlanders. Owner Gordon named them after the famed Scottish regiment, though the moniker also described the elevated terrain on which the park was built.

Hilltop, born of a shady political deal, was unwelcomed by Washington Heights residents, who were none too pleased by the prospect of thousands of baseball fans invading their quiet neighborhood. (Andrew Freedman is rumored to have stirred up neighborhood opposition — and Freedman was such a truculent fellow that maybe he caused this bit of trouble, too, just for laughs.) For its first three seasons, the park was almost an hour train ride from downtown New York, though by 1906 a new subway line reduced traveling time by half. In 1913, the New York team, now called the Yankees, moved into the Polo Grounds, home of the Giants. Two years later, "Colonel" Jacob Ruppert, heir to a brewing fortune, and the delightfully named Tillinghast L'Hommedieu Huston, an engineer and Spanish-American War veteran, bought the team. By 1920, the previously cellar-dwelling Yankees were doing very well on the field — thanks in part to a Baltimore orphan named Babe Ruth — and the Yankees were now too proud a franchise to share a ballpark with anyone else, especially as a mere tenant whose landlord was the National League's Giants. The Yankees were outdrawing the Giants and so the senior partner in this uneasy cohabitation suggested that the junior team might best hit the road. It was time to move — it was time to build an edifice fitting and proper to what would become the greatest — and most infamous — team in American sports.

This time, there would be no begging favors of Tammany Hall hacks. On February 6, 1921, the Yankees announced that they were moving across the Harlem River to a ten-acre parcel of land in the Bronx. Purchased for $675,000 from the estate of William Waldorf Astor, the late heir to the Astor family fortune, the property was close enough to the Polo Grounds that fans of the Giants could look out over the river and see a better team, the Yankees, led by demigods Ruth and Gehrig. Astor, who had died in 1919, would have been bemused by this usage of his property. A rich dilettante whose Republican Party connections had earned him a stint as ambassador to Italy under President Arthur, Astor fled the United States and became a British citizen, famously announcing that "America is not a fit place for a gentleman to live."[31] What Babe Ruth would have made of the expatriate's declaration can only be guessed.

Jacob Ruppert bought out Tillinghast L'Hommedieu Huston to become sole owner of the Yankees and the man most responsible for the new stadium. Colonel Ruppert, the "Prince of Beer" whose sobriquet came courtesy of the Ruppert Breweries upon which his family fortune was based, remarked, "Yankee Stadium was a mistake. Not my mistake, but the Giants' mistake."[32] This being New York, politics was inescapable: Ruppert's family were major contributors to Tammany Hall. In fact, the Colonel — the title was largely an honorific for his brief service in the National Guard, so he was no more martial than your typical Kentucky Colonel — served from 1899 to 1906 as a Tammany-backed Democratic congressman from the Upper East Side. In New York City, Major League Baseball is *always* political.

Giants manager John McGraw scoffed, "They are going up to Goatville. And before long they will be lost sight of. A New York team should be based on Manhattan island."[33] But this New York team was only a 15-minute subway ride from downtown Manhattan, and besides, it featured the greatest and most colorful player in the history of the game.

For $2.5 million, the White Construction Company of Madison Avenue, New York — with a little help from Babe Ruth — built Yankee Stadium, the gargantuan triple-decker park which opened for business on April 18, 1923, as the Yanks played the Boston Red Sox. Construction of what the *New York Times* predicted would be "the greatest baseball plant in the world" came in on time and on budget.[34] And *completely unsubsidized by New York's taxpayers.*

Before an Opening Day crowd of over 60,000, John Philip Sousa and his Seventh Regiment Band played the "Star-Spangled Banner," the Babe hit a home run (which he later said was his favorite of all the 714 homers he had hit), and the Red Sox lost — a perfect day for New Yorkers.

A Team Grows in Brooklyn

Over in Brooklyn, Dodgers' owner Charles Hercules Ebbets embarked on a series of moves that would eventually land his team in one of the most beloved ballparks in baseball history. Ebbets, an architect and New York Democrat who had served a year in the New York State Assembly and lost a try for the State Senate, was unhappy with the $7,500 annual rent the team was paying at the somewhat remote Eastern Park in the East New York neighborhood of Brooklyn. So for $60,000 he built Washington Park, near the site of the Revolutionary War Battle of Brooklyn and, perhaps more pertinently, close to a previous incarnation of Washington Park in which the Brooklyn Trolley Dodgers of the American Association had played. The Dodgers leased the property for 15 years beginning in 1898 — 15 sometimes malodorous years, for the nearby Gowanus Canal and assorted factories sent up a stench that disturbed even the most impervious olfactory senses of Brooklyn fans. The park was also on the small side, though the Dodgers of the early twentieth century were so bad that finding a place to put all the fans was seldom a pressing concern.

But the age of the wooden grandstand in the majors was coming to end, hastened by fires and fears of fire. Those owners who could afford it built spanking-new ballparks; those less flush renovated stadiums or toughed it out in venerable nineteenth-century ballparks. Charles Ebbets could afford it.

He bought land in the pungently named "Pigtown" section of Brooklyn for $200,000 and constructed upon it Ebbets Field, raising the money through bonds, banks, and contractors who came in as part-owners. How he pulled this off once again emphasizes the contrast between the government–baseball relationship of 100 years ago and that of the modern age. The land which Ebbets coveted in this less-than-desirable part of Brooklyn was owned, in bits and pieces, by more than forty people. This would pose an insuperable obstacle to the baseball owner of today: before you could say "condemnation" he would be demanding that his friends and allies and flunkies in the city government initiate eminent domain proceedings against these 40+ owners. The assertion would be made that these parcels of land suffered "blight," and that their highest public use would be achieved by bundling them up, seizing them after paying a "fair price" to the putative owners, and building, at public expense, a baseball stadium in which the wealthy owner might put his team.

The thought of stealing land from the proprietors of Pigtown never crossed Charles Ebbets' mind. Instead, he negotiated over three years the purchase of 1,200 parcels of land from these more than two score owners. True, he used something called the Pylon Construction Company as his front, for he feared that if an avaricious property owner caught wind of the purpose which he envisioned for the land that owner might, in a fit of rapacity, demand so high a price that the deal would fall apart like an error-plagued team in the seventh inning.

Buying land instead of stealing it via eminent domain could be a frustratingly slow process, though at least it enabled those involved to sleep the sleep of the just. As Michael Benson tells the story, "There were delays. One parcel of land could not be purchased as the owner couldn't be found.

Ebbets' people chased the landowner all over the world, and finally back to New Jersey. When they caught up with this traveling gentleman he proclaimed he'd all but forgotten his little piece of Brooklyn. Having no notion of the importance of the plot the man asked (and promptly received) a measly $500 for the land."[35]

Compared to piecing together the jigsaw of Pigtown land, building what the owner immodestly dubbed Ebbets Field was a can of corn. Ebbets went into a partnership with well-connected contractors (and brothers) Stephen and Edward McKeever. For $750,000, the troika built one of the baseball's truly charming parks, which even included an 80-foot rotunda with marble floor, coat-checkers, and curved seats whose design was borrowed from the opera house. Not that Brooklyn's fans would ever be mistaken for patrons of the fine arts.

The Dodgers were in many ways a glorious franchise. The intimacy of their park and the character of the borough of Brooklyn created a real bond between players and community. The team was the first to break the color barrier, as Jackie Robinson starred for the Dodgers for a decade beginning in 1947. That National League pennant-winning season also marked the high point of Brooklyn's attendance, as Dem Bums drew 1,807,576 into Ebbets Field.

Walter O'Malley, New York-born bankruptcy attorney for the team, became part-owner in 1944, along with the legendary Branch Rickey, the man who signed Jackie Robinson and broke Major League Baseball's color barrier. O'Malley became sole owner in 1950. He was not a man who lacked for self-esteem. As Lee Elihu Lowenfish writes, Walter O'Malley "liked to be called *the* O'Malley" and he "forbade mention of Rickey's name in his office for over two years."[36] He was not one to be bullied by New York politicians, even the imperious Robert Moses, the powerful and destructive "master builder" of the New York City bureaucracy.

The Crown Heights neighborhood, as the Dodgers home was called, had deteriorated, parking was limited, and "Ebbets Field now seemed an uninviting place in an increasingly unfamiliar neighborhood." The stadium had a sense of "claustrophobia created by the narrow seats and aisles" — though some thought this contributed to closeknittedness.[37] It was not terribly accessible, either, as only one subway line delivered commuters to the park. Ebbets Field was running out of time.

Walter O'Malley wanted land — taken from owners by urban renewal, or eminent domain — at the corner of Brooklyn's Atlantic and Flatbush Avenues on which to construct, privately, a $4 million stadium. (Perhaps even a domed stadium designed by Buckminster Fuller. Geodesic Field?) It would have been fed by three subway lines. "O'Malley was determined to build the new stadium with private capital, the first stadium so financed since Yankee Stadium was erected in 1923," wrote Neil J. Sullivan, author of the valuable *The Dodgers Move West*.[38] But he wanted stolen land on which to build it.

Robert Moses, then chairman of the city's Committee on Slum Clearance, rejected O'Malley's request for the land under Title I of the Federal Housing Act of 1949. O'Malley wanted the city to condemn the land he coveted and sell it to him. He would then build the new stadium with his own money. But Moses, the "Power Broker" of New York, in the title of Robert Caro's classic 1974 biography of the king of expressways and eminent domain, a man whose public works projects forcibly displaced half a million people, preferred a location in Queens. A typical bully, Moses baited O'Malley: "Then, if you don't get this particular site you'll pick up your marbles and leave town?"[39]

O'Malley was not acting in accordance with any pure notion of laissez-faire, or free enterprise. He wanted the city government to essentially steal the land upon which his stadium would be built. But he would pay for that land, and the ballpark in which the Brooklyn Dodgers would play would be financed by private means. Compared to George Steinbrenner, Joan Payson, and the others who have owned New York City teams since the Dodgers left town, Walter O'Malley was Milton Friedman, Ayn Rand, and Adam Smith rolled into one. And the funny thing was, as Sullivan observes, that New York offered O'Malley more than he actually wanted. Robert Moses, the unelected czar of the city, wanted to build him a stadium — O'Malley said no, he'd rather do it with his own money.

Robert Moses actually asked one good question, though given his history of stealing homes it must have been simply rhetorical. He wrote O'Malley on August 15, 1955, asking: "If you need only three and a half acres of land, if it is indeed distressed property, if you have a million dollars in the bank, if you have railroad easements, if you really want to stay in Brooklyn, why don't you buy the property at a private sale?"[40]

Good question, though there were a number of small businesses that O'Malley wished to displace, and rather than negotiate with each of them separately he'd much prefer the city government to swipe the land at one fell swoop via eminent domain. Charles Ebbets, by contrast, had purchased the land for his field from over forty separate owners between 1908 and 1911 — even if he tried, if not with complete success, to conceal his motives.

Brooklyn, meanwhile, countered O'Malley's Atlantic-Flatbush proposal with one of its own: a state-created Brooklyn Sports Center Authority would build a $30 million publicly owned stadium, financed with bonds. Since ultimate control would remain with government rather than in private hands, Moses came on board. The state legislature passed the necessary legislation and Governor Harriman signed it, but Mayor Wagner dragged his feet appointing members of the authority. "Those who wished to keep the Dodgers in Brooklyn faced the persistent opposition of Robert Moses, the hostility of officials from other boroughs, and the apparent indifference of the mayor," writes Sullivan.[41] New York City's Board of Estimate produced a study shot through with pessimism over the prospects of such a stadium turning anything close to a profit. Meanwhile, La-La Land came a'courting.

A continent away raged the "Battle of Chavez Ravine," which refers to a largely Mexican–American area just north of downtown Los Angeles. Three hundred and fifteen acres of Chavez Ravine were seized by the city of Los Angeles shortly after the enactment of the National Housing Act of 1949. The U.S. government had recently won a world war, and the prewar limits on just what that government could and should do were falling away. Providing housing was now within its portfolio; Chavez Ravine was 1 of 11 sites in Los Angeles targeted for housing projects. The architects were to be modernists Richard Neutra and Robert Alexander.

As Thomas S. Hines writes in "Housing, Baseball, and Creeping Socialism: The Battle of Chavez Ravine, Los Angeles: 1949–1959," Neutra and Alexander entered the project with a certain ambivalence. For the two architects, like most people who knew Chavez Ravine, "had to admit that the area was 'charming,' and that its people seemed to be 'happy' and well adjusted, with a rather intense feeling of pride in, and identification with, their community," which was Mexican and Roman Catholic in character. Its "lively street life" and "exquisite plantings" and ubiquitous window boxes bespoke a thriving, if not affluent, community, not at all the blighted slum which the National Housing Act's projects were supposedly designed to supplant. There existed in Chavez Ravine "espirit, vitality, and neighborhood identity," which baffled urban planners, who had been taught in their modern schools that such qualities did not spring up naturally in small neighborhoods but must be imposed, from the top down, by well-meaning planners with advanced degrees — planners who did not, and indeed would not dream of, living in the neighborhoods and projects they supervised. (Architect Neutra, however, insisted that Chavez Ravine was a "slum," despite its "human warmth and pleasantness," and he called its residents "Aztecs," not Mexicans.")[42]

In 1949, Chavez Ravine was home to about 1,100 families; the government planners intended to move another 3,360 families into the area, which was to resemble a dystopian futuristic city with "24 thirteen-story towers and 163 two-story structures." The history of this settlement — in fact, the history of Southern California — was irrelevant to the architects, who "included no allusions to Spanish colonial architecture" in their designs.[43] In other words, Chavez Ravine, as a friendly, homely, comfortable Mexican-Catholic hamlet, was to be destroyed. Neutra explained that "[h]owever romantic it may be to dream of retaining the present charm of rural backwardness...the area cannot be redeveloped with suburban bungalows. A realistic use of the site by any developer will require an urban housing solution."[44] Experts knew best!

The story, already a depressing tale of modern architecture and modern government's disdain for ordinary people, grows even grimmer. The people of Chavez Ravine were, in wartime fashion, "relocated." Their homes were condemned and seized by the city government. Even their church, despite a fight, was taken. Chavez Ravine was leveled, razed, eliminated — but before Neutra and Alexander could build their eyesores on this hallowed ground, the city council, sensitive to charges in then-conservative Los Angeles that public housing was socialistic, canceled the project. Chavez Ravine, empty and denuded of houses, people, and life, was up for grabs.

Enter Walter O'Malley and the Brooklyn Dodgers.

Los Angeles, which in the postwar years surpassed Chicago as the country's second-largest city, had been pursuing teams before the city fathers set their sights on Brooklyn. The City of Angels almost got the St. Louis Browns in 1941, but Pearl Harbor put baseball back on the bench. Still, the Second World War and the resultant explosion of defense spending inordinately benefitted Southern California and made Los Angeles an even more attractive city for team owners with roving eyes.

There had been some talk of trying to lure west the Washington Senators — first in war, first in peace, last in the American League, as the old joke had it — but Mayor Norris Poulson thought they were "tail-enders" and he wanted a top-flight franchise.[45]

The Dodgers were that franchise. Mayor Poulson seems to have approached O'Malley sometime around the 1956 World Series and invited his inspection of Chavez Ravine, now largely cleared of people and houses thanks to the National Housing Act's malevolent interventions. Only a handful of hardcore holdouts continued to resist eminent domain and eviction.

In February 1957, O'Malley took the significant step of buying the Los Angeles Angels of the Pacific Coast League and the team's park, the echoically named Wrigley Field, from Philip Wrigley of the Chicago Cubs. Events picked up speed, as they must when a team is to move 3,000 miles. O'Malley sold Ebbets Field to raise money for his new stadium. The Los Angeles City Council made its pitch for the Dodgers to come west. The city would give O'Malley 185 acres of Chavez Ravine and pay $2 million to grade its hills, level its mounds, build its access roads, and otherwise prepare the site for a baseball stadium. As Neil J. Sullivan writes, speaking of Los Angeles and Baltimore, which had recently attracted the Orioles, "Private enterprise was undoubtedly praised at numerous city functions in those communities; but if a dose of socialism was required to secure a major league franchise, then city officials were often willing to act like Marxists for the moment."[46] (O'Malley bluntly said that Chavez Ravine was "hill ground that would be of interest only to goats."[47] How quickly the displaced people who once lived in cleared "slums" and "blighted" neighborhoods are forgotten, especially by those who covet their land.)

Robert Moses made a counteroffer: a domed stadium at Flushing Meadows, Queens, which of course took the Brooklyn out of the Dodgers and the Dodgers out of Brooklyn. O'Malley was not interested. "The jig's up, Walter," the imperious Moses said, but the jig was not up for O'Malley, only for Brooklyn.[48] On June 26, 1957, O'Malley used the same expression, saying "the jig is up in Brooklyn."[49] The Dodgers were lame ducks. Next year in Los Angeles was the cry.

The National League's panjandrums had no objection to invading the West Coast, but they did not want a lone Los Angeles franchise holding down the fort. Two teams would be much better than one. In May 1957, the league unanimously approved a move by the Dodgers to the West Coast as long as another franchise made a similar move. And so Horace Stoneham, New York Giants owner, picked up stakes, though his brow has been spared the vials of scorn poured upon O'Malley. For the Giants were a poor relation in New York baseball, stuck in the Polo Grounds, largely forgotten by the city's media, which focused on the Yankees and the Dodgers.

"I had intended to go to Minneapolis," explained Stoneham, but O'Malley's move coastward dragged him to California instead.[50] Or, rather, Stoneham leapfrogged O'Malley, the former officially announcing his move in August 1957, the latter doing so weeks later.

Brooklyn's last game was a melancholy affair. On September 24, 1957, 6,702 fans watched the Dodgers beat the Pirates, 2–0. And then Walter

O'Malley, whom Brooklyn fans now regarded as something worse than a pirate, skipped town with his team in tow.

The Dodgers had won pennants in six of their final 11 years in Brooklyn, including a World Series championship in 1955. Unlike the Boston Braves and Philadelphia Athletics, Brooklyn was a relatively healthy franchise. In 1955, the Dodgers had led the league in net profit after taxes ($427,195). As Sullivan points out, "the other teams that changed cities in the 1950s [Boston Braves, St. Louis Browns, Philadelphia Athletics] were destitute franchises" piling up financial losses as rapidly as they piled up losses on the field. Brooklyn, by contrast, was "the only National League team to make money each year from 1952 to 1956."[51] (They even outearned the Yankees over that stretch.)

As Arthur T. Johnson of the University of Maryland-Baltimore County notes, the moves of Brooklyn and the New York Giants signified that "Financial distress was no longer a necessary justification for relocation."[52]

Yet, there was a flipside, too. By the end of the Dodgers run, Ebbets had a capacity of 32,000, but it was seldom filled. Attendance averaged only half that of capacity in the pennant years of 1955 and 1956, and in 1956 Brooklyn played seven (well-attended) games in the Jersey City stadium as a way of emphasizing O'Malley's seriousness. As Cary S. Henderson writes in "Los Angeles and the Dodger War, 1957–1962," Brooklyn's "attendance records were more indicative of a last-place team" than the front-runners they had been.[53] Attendance declined between 1953 and 1957, just as Brooklyn's population was declining. As Barry Jacobs writes, "over 235,000 largely middle-class whites left, most moving out to the suburbs."[54] The demographic writing was on the wall, or so it seemed to some. Moreover, Ebbets "had limited parking space and was near only one subway line."[55]

Neil J. Sullivan argues that because Brooklyn had forfeited its status as an independent city in 1898 and become just another borough of Greater New York, its political strength was diluted. Brooklyn was no longer one of America's largest and proudest cities. Rather, it was a second fiddle, a poor relation, to Manhattan, and as a subordinate borough in the city of Greater New York it needed the help of the other boroughs. In the matter of keeping the Dodgers home, "none was forthcoming."[56]

Interestingly, the legitimacy of the use of eminent domain was a key feature of the Brooklyn/Los Angeles equation, and New York City — which is, in the stereotype, the home of big government-wielding meddlesome socialists — well, here is how Sullivan puts it: "In New York, the point of view was that a baseball stadium did not merit the use of governmental authority to acquire needed land; in Los Angeles city and county officials believed a major league stadium fit the legal requirements for the use of Chavez Ravine."[57]

O'Malley may have wanted to stay in Brooklyn, but Robert Moses, who ruled Gotham City, would have none of it. Rather than seize land and sell it to O'Malley, so the owner could build with his own funds a stadium at Atlantic and Flatbush, Moses insisted upon a publicly subsidized stadium in Queens.

By contrast, Los Angeles simply gave O'Malley Chavez Ravine. Approved by a City Council vote of 11–3 on September 16, 1957, the city traded 315 acres of the condemned land for the minor-league ballpark and nine acres, which were the home of the Los Angeles Angels of the Pacific Coast League. And Los Angeles agreed to spend $2 million readying the site for Major League Baseball. It seems that "creeping socialism" was in the eye of the beholder.

Several lawsuits by irate taxpayers delayed the transfer of the land to the Dodgers, but the real hurdle facing the giveaway of Chavez Ravine was a June 3, 1958, referendum that was forced by petition-passing citizens who collected 85,000 signatures. The anti-giveaway group called itself the Citizens Committee to Save Chavez Ravine for the People, and while rumors circulated that John Smith, owner of the Pacific Coast League's San Diego Padres, was helping to fund the effort — the PCL sensed doom if the majors started spreading up and down the California coast — its popular appeal was undeniable.

The LA referendum jolted O'Malley: "I was completely unaware of the thing they called a referendum because they never had that in New York. Very few places have it. [Actually, quite a few do.]...Very peculiar. No boss."[58]

Mayor Poulson later confessed that the pro-stadium people ran a "scare campaign" during the referendum to fool the people into believing that it "was unalterably a yes-or-no vote for baseball."[59] If Chavez Ravine were not given to O'Malley, he would head right on back to Brooklyn, or some other non-Los Angeles site. This was exceedingly improbable, but it worked.

The pro-giveaway forces, grouped under the comedian Joe E. Brown's Committee on Yea for Baseball, sponsored a telethon featuring the likes of George Burns, Groucho Marx, Ronald Reagan, Debbie Reynolds, and Jack Benny. Who could resist? In the end, the referendum passed by fewer than 25,000 votes out of over 666,000 cast: 345,435–321,142. Hardly an overwhelming endorsement of the gift of the stolen land of Chavez Ravine, but the Dodgers would take it.

By this time, of course, the Brooklyn Dodgers had become the Los Angeles Dodgers. They were playing in the football-friendly Los Angeles

Coliseum and losing plenty of games en route to a seventh-place finish in the National League, but they were smashing attendance records. Los Angeles was a fertile market for Major League Baseball. The Dodgers played to huge crowds in the Coliseum: they drew an incredible 78,672 for their home opener, and despite a mediocre team they drew 1,845,268 in that inaugural season, which was better than Brooklyn's highest attendance. The Coliseum, which was not a good baseball park, had a left field foul pole only 251 feet away — a good Little Leaguer could hit it that far, though a 40-foot high screen blew back cheap homeruns. Even worse, it sold no beer! Yet, Los Angelenos thirsted for major league sports, and while O'Malley's betrayal of Brooklyn would go down in American sporting history as an act of villainy, the fact that he (1) built Dodger Stadium on his own dime (on land that had been acquired in a despicable manner, to be sure); and (2) brought the majors to a city that clearly deserved a team complicates the Manichean tale somewhat.

The Los Angeles referendum having passed, the only hope for the anti-giveaway forces was the courts. Despite a promising lower court ruling, the California Supreme Court upheld the city's use of public funds to improve the Chavez Ravine property for the Dodgers. All that was left was to evict the last of the holdouts. And here things got sticky.

With TV cameras rolling, May 8, 1959, turned into what came to be called "The Battle of Chavez Ravine." As Cary S. Henderson writes, "The door to the home of two elderly ladies had to be battered down because they had locked it and refused entry to the police."[60]

Even more tenacious was the Arechiga family. On May 8, Mrs. Avrana Arechiga, 68, hurled rocks at the police as they sought to evict the family from the two "run-down houses" in which the extended family lived. Two of Mrs. Arechiga's grown daughters were "physically carried out of the house." Grandchildren were "wailing hysterically." The family attracted widespread sympathy, as outraged citizens called the acts of the city "communism" and defended the property rights of this Mexican–American family that was refusing to bow under enormous pressure from the establishment. They stayed put for ten more days, as the newspapers, which were vigorously pro-Dodgers, argued that since the family owned other houses as rental properties they should just give up the ghost. Mayor Poulson even called them "ham actors...bleeding insincere tears," and pro-stadium forces sought to dissipate public sympathy for them as underdogs being displaced by greedy villains.[61]

The Arechigas lost. The system won, and Chavez Ravine was delivered to Walter O'Malley, along with $2 million in improvements to the land. On September 17, 1959, ground was broken for Dodger Stadium, which would open in April 1962. In their second year in Los Angeles, the Dodgers won the World Series in 1959, with the Coliseum as their home. And though the story of how Chavez Ravine was cleared of people and then ceded to Walter O'Malley is a rather squalid tale of government force on behalf of powerful interests and against ordinary people, there is another side to the story. For Walter O'Malley and the Dodgers did build their own stadium — a private facility, constructed with $23 million that did not come from government handouts or sales taxes or levies on automobile renters or out-of-towners staying at hotels.

O'Malley, however disloyal to Brooklyn, was not as egregious a welfare case as most other owners. As he said in August 1955: "To clear up any misconceptions, I would like to make it plain that we are not going into this thing with our hats in our hands. We are — and have been for some time — ready, willing and able to purchase the land and pay the costs of building a new stadium for the Dodgers."[62]

Although the 55,000-seat Dodger Stadium was built with private funds, the City of Los Angeles did spend $2 million grading the site, and the county public works department spent $2.7 million putting in roads to the site. It was hardly an exercise in pure free enterprise, but given the time and the tenor of that time, it was as close to a private undertaking as one could find in Major League Baseball. Dean V. Baim, in a study of 14 subsidized stadiums, found that the only one that had returned its public investment was Dodger Stadium, which had paid back to the municipality more in property taxes than it had received in subsidized infrastructure improvements.[63]

As for the betrayed borough of Brooklyn, Barry Jacobs, writing in the journal *New York Affairs*, concludes that although the "psychological impact of losing Brooklyn's leading light has been great," the "economic impact of the team's 1957 move to Los Angeles on the borough. . .was minimal."[64]

Ebbets Field was demolished in 1960; when it comes to real estate, the boroughs of New York are not the most sentimental places. The "restaurants, shops, and other once-thriving businesses" that had surrounded Ebbets Field when the Dodgers called it home soon shut their doors.[65] Ebbets Field Houses, a housing project, went up on the ashes and ruins of the old ballpark.

Ron Miller, a historian of the borough at Brooklyn College, told Barry Jacobs that the effect of the departure of the Dodgers was "symbolically important" but not a significant factor in Brooklyn's subsequent economic decline — a decline that was accelerating before the Dodgers had traded in subway stops for the freeways of LA.[66]

If Charles Ebbets' Brooklyn Dodgers had once been a model of a private baseball club pursuing profit by free enterprise means — an imperfect model, to be sure, given the arguably surreptitious method in which they obtained the land for Ebbets Field — the Los Angeles Dodgers were to serve as a far lonelier model half a century later, in very different times. Neil Sullivan concludes, "The Dodgers and Los Angeles have thrived as a model association of community and ballclub. Citizens and the franchise both recognize that ball clubs are businesses, not wards of the state, and that they should be responsible for providing their own ballparks with limited government assistance."[67]

Would that more teams followed this model.

The Classic Parks

Before the War to End All Wars, two ballparks went up that stand today as monuments to the genius of early baseball architecture and the willingness of owners to pay their own way. It is no coincidence that fans of the Boston Red Sox and Chicago Cubs are among the most sentimental in all of baseball.

The Red Sox, in their previous incarnation as the Boston Pilgrims, had been playing at the Huntington Avenue Grounds, a much less poetic name than Fenway Park, and also a less secure site. As Robert T. Bluthardt notes, at Huntington the Red Sox "rented the land from the Boston Elevated Railway Company, and the Boston Elevated could give the team a year's notice by paying them $3,000."[68] The Huntington Avenue Grounds, site of the first World Series game played in 1903, did have an inviting right-field wall only 280 feet from home plate, but even it was no match for the "Green Monster" which the Red Sox erected in Fenway's left field in 1936.

Built in a marshy and theretofore undeveloped section of Boston known as the Fens, served by streetcar lines, its exterior a suitable red brick, in keeping with Puritan modesty, Fenway Park was built for a grand total of $650,000 in 1912. Moreover, it was quirky and asymmetrical, for owner John I. Taylor hadn't the power (or desire) to seize whatever nearby real estate he might have desired. How much easier it would have been to simply condemn the whole neighborhood and steal it! And how much poorer the history of the Red Sox would be if they had been condemned to play in a government-built cookie-cutter ballpark.

Fenway opened for business in 1912; two years later, Wrigley Field, nee Weeghman Park, was privately built by Charlie Weeghman, owner of the stadium's first tenant, the Chicago Whales of the upstart Federal League. The Whales lasted but two years, whereupon Weeghman and a syndicate including William Wrigley bought the National League Chicago Cubs and moved them to what would eventually become Wrigley Field.

Wrigley Field is the second-oldest park in baseball, younger only than Fenway Park. Like Fenway, it is beloved for its idiosyncrasies, ranging from the ivy-covered outfield walls to the "bleacher bums."

Fenway and Wrigley Field are the only 2 of the 13 ballparks built in the golden age of "new park" construction between 1909 and 1915 still extant. These parks, said a writer of purplish prose in the *Philadelphia Inquirer*, turned a "fresh and scintillating page in the annals of the horsehead sphere."[69]

They also moved fans farther from the playing field, plugged the literal and sometimes metaphorical knotholes in the outfield fences, and "standardized and depersonalized the sport," argues historian Bluthardt.[70] This is a useful reminder that the things we are nostalgic for often replaced things that our ancestors were nostalgic for. In the matter of baseball parks, however, the array of ballparks built in the prewar teens were — making allowances for the usual grubby politicking and streetcar-line-manipulating — done through the private sector. The ballparks that replaced the grand old parks of the teens were thrown up, so to speak, by the government, except for Fenway and Wrigley, which endure.

A Capital Sport

In the heart of the world of politics, Washington, DC, the hapless Senators eked out an existence in the netherworld of the standings under two generations of the Griffith family. Clark Griffith, "the Old Fox," was a screwball and spitball pitcher at the turn of the twentieth century who in short order became manager, president, and then owner of the Washington Senators, a team he guided in various capacities from 1920 to 1955. Washington — the team, according to some wags, of the nation's "work-free drug place" — played at Griffith Stadium, named for Clark in 1920, though baseball had been played on this Georgia Avenue site since 1891.

Not being a modern owner, Clark Griffith had the quaintly old-fashioned idea that in putting together parcels of land on which to build a ballpark, one should negotiate — voluntarily — with the owners of that land. He did so, and since not everyone was willing to put a price tag on his home,

"Griffith Stadium's center field wall respectfully skirted the houses of several holdouts," as baseball historian Robert F. Bluthardt writes.[71]

Fittingly, the Senators had a political instinct. As Michael Benson relates, "The Senators used to give a gold season pass to the United States president Teddy Roosevelt, the first to receive this gift. Then, in a great public relations move, they got President William Howard Taft to throw out the first ball, stamping government approval on the national game" in the 1910 season opener.[72]

Alas, Teddy Roosevelt was a big-game hunter but not a big-game follower. His daughter, the irrepressible Alice Roosevelt Longworth, said that her father thought of baseball as a "molly-coddle game." Taft, by contrast, actually liked baseball. Historian George C. Rable, in his study "Patriotism, Platitudes and Politics: Baseball and the American Presidency," calls him "the first genuine presidential fan." He followed not only the Senators, whose ace Walter Johnson had captivated the nation's capital, but the Chicago Cubs, who "even installed a special box to accommodate the President's large frame."[73]

But the Senators were not so political as to coax federal funding for Griffith Stadium, which when Taft and TR were taking in games was called National Park. Clark Griffith, who was not an independently wealthy man, pinched pennies, but he did pay for improvements to the park he owned. By renting it out to Negro League teams, the NFL Redskins, and local college squads, and by controlling concessions, Griffith got by. When his son Calvin fled D.C. for the colder climes of Minnesota in 1961, and the American League awarded an expansion franchise to Washington to replace the now-Twins, Griffith Stadium gave way to what Benson calls "the first of the cookie-cutters, and the only federally owned ballpark ever used in the majors" — D.C. Stadium, which after Senator Robert Kennedy's assassination was renamed RFK Stadium.[74] How apt that the only stadium built and owned by the feds was widely derided as charmless and undistinguished. Somewhere, Clark Griffith was grimacing.

The Orioles Take Flight

The mid-1950s brought the big leagues to Baltimore. In 1953, the Dallas Colts of the NFL moved east, and the next year Bill Veeck's St. Louis Browns of baseball's American League followed suit. Both teams would play in Memorial Stadium, a city-owned postwar single-level minor-league ballpark (for the Orioles of the AAA International League) that had been constructed in 1950 with eventual expansion and the addition of a second tier in mind. The city responded with $7.5 million in taxpayer-funded improvements in 1953–1954. Dubbed "The Dowager of 33rd Street," Memorial Stadium exemplified the transition from private to public ballparks that would gain steam in the coming decade. Harry Von der Horst had rebuilt Union Park with money from his brewery in 1894, but Baltimore enjoyed no such brewing benefaction when the Orioles reentered Major League Baseball in 1954.

The St. Louis Browns' move to Baltimore was not without hurdles. Bill Veeck was not so cordially detested by the other owners, as well as by Commissioner Ford Frick. Veeck had wanted to move the Browns in 1953, but the New York Yankees objected, putting the kibosh on it. (Transfer of a franchise required the unanimous agreement of the league's owners; prior to 1952, it had required unanimous consent from both leagues.) Among those who tut-tutted Veeck for wanting to abandon St. Louis was Roy Mack, son of Connie and co-owner of the Philadelphia Athletics, or A's for short. The A's were a charter member of the American League; under manager Connie Mack, the team was a pre-World War I powerhouse, playing its games beginning in 1909 in the aforementioned Shibe Park, which the A's built on their own dime. The A's resurged in the 1920s, but the Great Depression was mirrored at Shibe Park by a 20-year downturn in the fortunes of the Philadelphia Athletics.

When Veeck wanted league permission to go east to Baltimore, Roy Mack declared loudly in an owners' meeting, "I don't see why you want to leave St. Louis. We'd never leave Philadelphia." Never say never, Roy.

As Veeck describes it in his autobiography, *Veeck — As in Wreck,* "I broke back in to tell him that Philadelphia would most assuredly be the next franchise to take wing. 'Oh no,' he said. 'No, never.' He delivered a moving speech about the glorious traditions of Philadelphia baseball, interspersed with a few kind words about civic loyalty and eternal ties. When a baseball owner talks about civic loyalty *to other club owners,* you know the fix is in."[75]

In October 1954, Mack & Co. sold out to Arnold Johnson, who took the team to play the 1955 season in Kansas City, where it played at Municipal Stadium, which had been renovated for Major League Baseball by the municipality of Kansas City, which is to say its taxpayers. Once again, a team playing in a privately built and owned stadium had followed the money to a ballpark built with public funds. The rush was on.

Meanwhile, the St. Louis Browns did move to Baltimore, though without Bill Veeck, who had been pressured by MLB to sell the team. (He came back later, with a vengeance, in Chicago with the White Sox.)

Alas for Baltimore, the mere presence of a big league ballclub did not transform a declining city. The Orioles were a subpar team, at first, as the Browns had been, and whatever excitement their presence brought to the city in 1954 wore off quickly. The team drew poorly. And the local shopkeepers were in for a rude surprise. As historian James Edward Miller writes in his study of Memorial Stadium for the *Maryland Historical Magazine*, "The initial stages of the city-team relationship were. . .marked by a slowly dawning realization among many members of the business elite that major league sports were an expensive drain on the public purse and that they failed to return any real economic benefits. Although the team provided a fairly large number of low-paying, part-time jobs, it created very few full-time employment opportunities. Local merchants, who had expected to cash in on a steady flow of customers into their shops, discovered that most fans headed directly to the stadium and spent their dollars at concession stands. Restaurants, hotels, newspapers, public transportation, and even taverns enjoyed only marginal gains in income."[76]

The city lost three quarters of a million dollars over the Orioles' first three years at Memorial Stadium, due to spotty attendance and a lease that gave the ballclub the major cut from parking and concessions. But as Miller explains, Baltimore Mayor Thomas D'Alesandro Jr. — whose daughter Nancy Pelosi would go on to serve as Speaker of the U.S. House of Representatives — was fully committed to spending as many taxpayer dollars as were needed to keep the Orioles happy. The city pumped another $1.25 million into Memorial Stadium in 1957. It would be followed by further infusions of cash, though by the mid-1960s the football Colts' colorful owner, Carroll Rosenbloom, a man with more gambling connections than Bugsy Siegel, was campaigning for a new $100 million stadium. (You can't entirely blame Rosenbloom for being peeved, given that the lease gave the Orioles a piece of the Colts' concessions revenue but not vice versa.)

Howard Cosell, if you will recall, had complete and total faith that the Colts would never move out of Baltimore. He was wrong. The Orioles did stay, however, and the ballpark that kept them anchored to the city is considered a model of the publicly funded sports venue. Although even it is no bargain for Baltimore, as seen in the next chapter.

Westward Ho!

The St. Louis Browns' move east was an anomaly; professional sports franchises began to shift westward in the 1950s, responding to population trends and improvements in transportation.

The movements of St. Louis (to Baltimore), the Boston Braves (to Milwaukee), and Philadelphia (to Kansas City) in the 1950s caused minimal psychic trauma, since each of the former cities still had a major league team, and it was generally agreed that Baltimore, Milwaukee, and Kansas City "deserved" teams. They certainly supported the new teams, to varying degrees. Los Angeles, which had snatched the football Rams from Cleveland in 1947, was ready for the Dodgers. Just as the population of the United States was shifting westward, so were the teams: in barely five years, a quarter of the teams in Major League Baseball up and moved.

In some respects, these moves were long overdue. As the irrepressible bad boy of baseball owners, Bill Veeck, cracked in his 1962 autobiography *Veeck — As in Wreck*, "If baseball owners ran Congress, Kansas and Nebraska would still be trying to get into the union."[77]

The West Coast had experienced extraordinary growth in the 1940s, much of it fueled by government spending on the war and the burgeoning defense industry. As baseball historian Lee Elihu Lowenfish writes in "A Tale of Many Cities: The Westward Expansion of Major League Baseball in the 1950s," in the decade of the 1940s "the growth rate of the entire Pacific Coast region was 24% compared to a national growth rate of only 2%."[78] And yet the majors had a grand total of zero teams located west of St. Louis. The Gateway to the West was as far west as Major League Baseball dared to venture. (There had been talk of a team moving to Los Angeles in 1941, as Lowenfish points out, but Pearl Harbor put an end to that.)

It is not that no market for baseball existed in the West. The Pacific Coast League was the jewel of the minors — in fact, it was clearly superior to the other minor leagues. With teams in Los Angeles, Hollywood, Oakland, San Francisco, Portland, Sacramento, Seattle, and San Diego, the PCL drew well and was able to attract excellent ballplayers, some of whom actually turned down major league opportunities in favor of the warm weather and better-than-minor-league rates of pay of the Pacific Coast League.

In 1946, the President of the Pacific Coast League, the amusingly named Clarence "Pants" Rowland (as a boy in Dubuque, Iowa, he wore his dad's overalls while madly running the bases, thus the nickname), petitioned the majors for a special status for the PCL as "The Pacific Major League."[79] The majors laughed at Rowland's presumption — didn't Pants know that there were two, and only two, divinely ordained major leagues? In any event, the minors, the major–minor PCL included, swooned in the 1950s, largely as a result of television, and the proposal was shelved. But there were still several PCL cities that had demonstrated that they could support a baseball team with major league-like attendance figures, and as air travel became more routine and less expensive, expansion or relocation was, well, in the air.

Unless and until a federal sports commission of the sort desired by Howard Cosell and the late Senator Cook comes into being, approval or disapproval of franchise moves rests now, as it always has, with the other owners in a league. Remarkably, there were no franchise moves among the 16 baseball teams of the American League and National League for a full half-century (1903–1953). This stability coincided with the emergence of Major League Baseball as the nation's most popular professional sport. It is probably not coincidental that during this time, any franchise move required unanimous consent from the other owners in the pertinent league, as well as approval by a majority of owners in the other league. After 1952, the other league lost any say in franchise movement (unless the proposed relocation encroached upon the territory of an existing franchise) and only a 3/4 (later modified to 2/3) vote was required within the affected league. After that, the rush was on.[80] (It should be noted, too, that in the 1950s franchises moved occasionally in the NFL and frequently in the NBA, the new pro basketball league. Seven franchises disappeared in the NBA in 1951 alone, and the decade also saw Milwaukee move to St. Louis in 1956 and Fort Wayne move to Detroit and Rochester relocate to Cincinnati in 1958. The 1960s saw even greater shifts, especially after the American Basketball Association was founded in 1968.)

As James Quirk of the California Institute of Technology, one of the first economists to study the economics of baseball, noted in a 1973–1974 paper, "franchise moves emerge as a device to permit small city franchises to capture short-run profits in a new market." An owner spots "higher profit potential" in another city.[81] The problem, of course, is that for almost 60 years now, that "profit potential" has been based in part upon municipal subsidies.

And the forerunner of this pernicious trend was the largest city in Wisconsin. As Dean V. Baim writes, "Milwaukee was the first city to use a city-built facility to lure a team from one city to another."[82] The Brewery City has a lot to answer for — as if *Laverne and Shirley* wasn't enough!

The first team to pack up and leave for greener pastures was the Boston Braves, an original National League franchise that had established a long record of futility and spotty attendance. This move, while it has received nowhere near the retrospective attention given the shift of the Brooklyn Dodgers to the eternal sunshine of Los Angeles, was of seminal importance: not only did it launch a mini-flood of baseball franchise moves in the supposedly placid 1950s, but it also created a precedent for a team-less city to lure a vagabond franchise with the promise of a taxpayer-built palace in which to play.

In 1952, attendance at Boston Braves games dipped to 282,000, and owner Lou Perini, who once had pledged "never" to move the team, was beginning to reevaluate, in a Clintonian manner, the meaning of the word "never."[83] The Braves had played to small crowds and big financial losses in Boston — more than $1 million over the team's final three seasons — so no one could fault the team for moving.

In Boston, the team played in Braves Field, which had been constructed in 1915 on the dime of owner James Gaffney. Built shortly after the rival Red Sox put up Fenway Park, Braves Field was also a park with real character, so spacious that its outfield might be mistaken for the interstellar void. At one point, dead center was almost 500 feet from home plate, making a home run a virtual impossibility. Yet ballplayers adapt, and Braves Field became known for inside-the-park home runs, among the most exciting plays in baseball.

Even after the fences were moved in, the home runs tended to be hit by opposing batters. The Braves were often bad, and the fact that Lou Perini also owned the Milwaukee Brewers of the AAA American Association incised the writing on the wall.

Milwaukee was being eyed not only by Perini but also by the rebel Bill Veeck, whose Cleveland Indians had won the World Series in 1948 but whose next team, the St. Louis Browns, were, like the Braves, very much a second-rate team in a two-team city. Veeck offered Perini $750,000 for the territorial rights to Milwaukee, but Perini said no — much to the dismay of the civic boosters of the Greater Milwaukee Committee and the Milwaukee Association of Commerce. County of Milwaukee taxpayers, and the construction firms they subsidized to the tune of $5.768 million, were building (on the site of an old garbage dump) Milwaukee County Stadium in 1951–1952, ostensibly for the minor-league club but really for the major league team the city hoped to attract. Capacity was slightly over 28,000, but within a year of entering the National League it was expanded by another 15,000 seats. Esthetically? Well, as ballparks.com puts it, it seemed like "an old park without an old park's character."

Lou Perini had not grown rich by running from opportunity. In March 1953, the National League owners unanimously approved the move of the Braves to Milwaukee. They opened the season one month later, and as Lee Elihu Lowenfish writes, "By May 20, [1953,] the Braves had matched the entire 1952 Boston attendance en route to an impressive season attendance of 1,800,000."[84] With stadium expansion the next year, attendance exceeded two million, and the Braves had gone from being afterthoughts in Boston to the object of what star pitcher Warren Spahn called "the biggest and most worshipful following in the majors." In the words of one ecstatic fan in 1953, "This is the greatest thing that has happened to Milwaukee since beer."[85]

A World Series soon followed in 1957, and beer and baseball cheerfully and profitably coexisted in Milwaukee. Hank Aaron was hitting them out of a park far friendlier to the home run than Braves Field had been. The team declined by the early 1960s, and so did attendance. By 1962, it had dropped to 766,921, which to Lou Perini meant only one thing: SELL. His profits in Milwaukee had totaled $7.5 million, thanks in part to not having to build and maintain a stadium, and in 1962 he pocketed a cool $6.2 million in a sale to a party of Chicago businessmen. "[T]he sum total of their knowledge about baseball is zero," said Bill Veeck of the new owners.[86] Milwaukeeans did not respond well to the team being owned by Chicagoans. Nevertheless, as Glen Gendzel points out, over the course of their lifetime the Braves "averaged a phenomenal 94.4 tickets sold per 100 residents," or far, far more than the average for other National League cities (22.2 tickets per 100 residents) and American League cities (20.7 tickets per 100 residents).[87]

Yet something other than Hank Aaron homeruns was in the air. There was something new about these 1950 franchise moves — a newness that went beyond the mere fact of a change of venue. Teams, once considered fixtures in their cities, were now footloose if not fancy-free. And cities were footing the bill. Before Milwaukee County Stadium, the only publicly funded ballparks and stadiums in the majors or the NFL were Los Angeles Coliseum, Soldier Field in Chicago, and Municipal Stadium in Cleveland, and all three of those had been conceived, in part, as a means of attracting the Olympic Games.

Writing in *The American City* in November 1957, before the Dodgers had moved but after San Francisco, Baltimore, Milwaukee, and Kansas City had gotten relocated teams, Douglas S. Powell observed that "baseball, the great national pastime, is rapidly becoming an official municipal pastime."[88]

He went through the list. San Francisco had flashed $5 million — real money in those days, to borrow Everett Dirksen's phrase — to lure the

Giants from Manhattan to the swirling winds of what became Candlestick Park. In fact, this $5 million subsidy was approved by voters in November 1954, in a kind of "offer to build it and they will come" gesture. They did indeed come.

In a reversal of what one might regard as the natural political order of things, East Coast politicians were more frugal than those of the West Coast or Midwest. New York City Mayor Robert Wagner, responding to threats of relocated teams, said that Gotham would not be "blackmailed into helping the Dodgers or Giants financially." Philadelphia Mayor Joseph Clark would not hear of the City of Brotherly Love offering taxpayers' dollars to keep the Athletics from moving to Kansas City; rather, Mayor Clark urged local businessmen to "pick up the ball" and save the team through a "community chest kind of drive."[89] This faith in voluntary action, though warming the hearts of the fiscally responsible, was at odds with the tenor of those post-New Deal times, when politicians exhibited a naïve faith in the power of government to perform all manner of good deeds. Mayors Wagner and Clark sounded like anachronisms — like relics of the age of Ebbets Field and Shibe Park. The day of big government was upon Americans, and privately financed baseball parks didn't fit in.

Voters and governmental bodies in Milwaukee, Kansas City, and Baltimore also approved bond issues to build or refurbish stadiums to attract major league vagabonds. In Milwaukee, the County Board of Supervisors dedicated $6.65 million to build County Stadium, which came to house the Braves and later, for a time, the Brewers. In Kansas City, the public voted $3.25 million in bond issues to redo circa 1923 Municipal Stadium, once the site of a frog pond, later home of the legendary Kansas City Monarchs of the Negro American League, and from 1955 to 1967 home of the bumbling Kansas City Athletics. (It was also home to the Kansas City Royals from 1969 until they moved into Royals Stadium in 1973.) Baltimore voters three times, between 1947 and 1953, voted to spend $6 million building and bulking up the stadium that lured the St. Louis Browns to the Charm City.

The American City's Powell noted that the municipalities had in each case negotiated lopsidedly generous leases with the new teams. Partly as a result, the cities posted large average losses over the three years preceding 1957: Baltimore (average loss $241,781), Kansas City (average loss $183,267), and Milwaukee (average loss $426,855) were certainly paying for the privilege of hosting a major league team. Yet, city fathers tended to think it was worth it. As a Milwaukeean boasted, "The team is the symbol that we've become a big city in every way." An expensive symbol it was.

Still, the mid-1950s, an era in which the federal government was launching such massive projects as the Interstate Highway System and the space program, as well as funding a Cold War, were a welcoming age for new government ventures. Even after surveying the money-losing experiences of the new major league cities, analyst Powell wrote in 1957 that "[i]n the long run it may well be that spending city tax money to attract to or even keep a big league team in a city is merely another part of the package of city aids including industrial development [and] offstreet parking aimed to reviving the waning life of today's metropolises."[90] He might also have mentioned urban renewal, or slum clearance, as an example of such resuscitative measures. That city-sponsored development, the profusion of parking lots, and urban renewal helped to devastate rather than renew urban centers would not become clear for another decade or more, but to anyone with a modicum of skepticism about the salvific properties of government aid, the writing should have been on the wall.

As Neil J. Sullivan wrote in "Major League Baseball and American Cities: A Strategy for Playing the Stadium Game," the worst thing to come out of Wisconsin in the 1950s was not red-baiting Senator Joe McCarthy but the example set by the city of Milwaukee in using taxpayer dollars to lure the Braves to the Dairy State.[91]

The Minors and Negro Leagues

Government support of minor league teams was also rare in the early years of baseball. Funds were often raised by subscription from enthusiastic fans and local businessmen. Typically, as Harold Seymour writes, "Fans in Paragould, Arkansas...bought the team's uniforms in 1910. Newark fans contributed $125,000 needed to complete the club's stadium in 1926 and actually oversubscribed the fund by $5,920."[92] The teams, often owned by groups of local merchants and men of prominence, supported themselves from ticket sales, advertising, and the sale of their best ballplayers to major league clubs. Few got rich in minor league baseball — the community of fans was the entity that really profited, as they watched good baseball in an atmosphere of fresh air and fraternity. (By the turn of the twenty-first century, the city of New York would spend $76 million to build a minor-league Single A baseball park for the Staten Island affiliate of the Yankees.)

The Works Progress Administration, the oft admired and just as often derided make-work program of President Franklin D. Roosevelt's New Deal, put men to work building baseball stadiums in small cities and towns, among its many other projects. From Aberdeen, South Dakota, to Raleigh, North Carolina, the WPA constructed no-frills parks in which minor league and local amateur teams played from the 1930s until, in many cases, the 1980s or 1990s. (Many of the surviving WPA stadiums were drastically renovated or knocked down in those later decades at the behest of Major League Baseball, which ordered their minor-league affiliates to spruce things up — often at the expense of state and local taxpayers.)

"By 1937 the WPA and other government agencies had built 3,600 baseball and 8,800 softball fields," and as a result, Harold Seymour noted, youth baseball tended to move from the informal sandlot ball to a more organized and adult-overseen game.[93] The quality of the fields improved, without doubt, though maybe a certain spirit was lost in transferring control of the game from high-spirited boys (who knew the rules by heart, and who somehow made it through argued balls and strikes and disputed calls at first) to supervisory adults.

Even the Negro Leagues, whose very existence was a reminder of the painful reality of Jim Crow, were occasionally cut in on the action. As Michael Benson writes in his capsule history of Bacharach Park in Atlantic City, New Jersey, which hosted the Bacharach Giants, an original franchise of the Eastern Colored League in 1923, "Mayor [Harry] Bacharach of Atlantic City went to see a black ball game in Jacksonville, and was so impressed by one team's skills that he hired them — actually, he put them on the city payroll — and built a ballpark for them to play in, naming everything he could for himself."[94]

This act of chutzpah would have made Boss Tweed sit up and notice. Mayor Bacharach apparently had some of Tweed's ethical laxity, too. He was tried for election fraud stemming from the 1910 race, though it didn't keep him from serving as mayor for five terms. When he died in 1947, *Time* eulogized him as the "longtime 'No. 1 booster' of Atlantic City...An ardent publicity-grabber (he once carried on the city's business in an amusement-pier office flanked by an educated chimpanzee and a half-man-half-woman), he nonetheless worked noisily at keeping his resort free of known thugs and 'undesirables.'"[95]

At least the scoundrel was upfront about everything. Naming the team and the stadium in which they play after yourself — while spending the money of the people of Atlantic City on such self-aggrandizement — is almost breathtaking in its brazenness. Even the "thugs" and "undesirables" must have admired Harry Bacharach for that act.

Football Takes the Field

Switching from the pastoral game of baseball to the violent sport that would oust it from its perch as the favorite American game, we find no shortage of subsidized playing fields. The earliest government-funded stadiums generally were built to attract the Olympics or hold massive crowds for boxing matches or college football games.

San Diego was among the first cities to build a municipal stadium in 1914: a 30,000-seat oval structure whose $150,000 cost was paid for by taxes and admission fees. It hosted baseball and football games and, in a throwback to *Ben Hur*, chariot races.[96]

Stadiums built with local government money before or early in the Great Depression included the Rose Bowl (1922), the Los Angeles Coliseum (1923), Soldier Field in Chicago (1924), and Municipal Stadium in Cleveland (1930).

Soldier Field was intended by its architects to be spectacular, a "classic temple" befitting the rising city of Chicago. The idea was given birth in 1919, as the First World War — whose dead were to be memorialized by the stadium's name — was about to give way to the Roaring Twenties. The South Park Commission, whose purview included Chicago's Grant Park, chose the architectural firm of Holabird and Roche to design a stadium to grace the southern end of Grant Park. The architects envisioned a "U-shaped amphitheater" whose influences included the Parthenon, Greek temples, and other "authoritative Classic Greek sources."[97]

The immodest dream behind Soldier Field was stated in grand terms: "The erection of this mammoth structure will serve to stimulate a spirit of good sportsmanship, developing contests, and physical superiority that will assure new records."

Alas, as Perry Duis and Glen E. Holt note in "The Classic Problem of Solider Field," the "classic temple" of the Second City was fraught with problems from the start. To fit the stadium for multiple uses, the architects elongated it to a considerable degree. Combined with the sloped seating, the effect was such, as a Chicago *Sun-Times* writer complained in 1958, that "you need binoculars in the rear pews, and the ends of the stadium are in the next county."

In 1920, voters endorsed a $2.5 million bond issue. Unfortunately, the low bid on Soldier Field came in at $4.35 million. Subsequent bond issuances were necessary, and as costs rose — Chicago being Chicago, costs did rise — the stench of political corruption emanated from the construction.

Edward J. Kelly, at various times president of the South Park Commission and Mayor of Chicago, was widely believed to have arranged for kickbacks for himself and the Democratic Party. Good-government groups were outraged — though not surprised. After all, it was Chicago.

Soldier Field was built, and it served its purpose capably for several decades, including as home of the NFL Bears. Yet, its construction cost of $8 million — from the public purse — was extraordinarily high for that time. Duis and Holt observe, "The City of Los Angeles, meanwhile, was building a larger civic stadium at a cost of $1.7 million — a little more than one fifth of what Chicago spent."[98]

The Los Angeles Coliseum was the expression of that city's abundant post-World War I chamber-of-commerce boosterism. Los Angeles, long regarded as distinctly inferior to San Francisco in the hierarchy of California cities, was beginning its twentieth-century flourish. The city was growing; word had spread that it was the closest thing to Paradise on the North American continent.

LA's boosters dreamed big. They set their sights on attracting the Summer Olympics to their town. These boosters had as their base the Community Development Association (CDA), which was composed of "the leading publishers, attorneys, bankers, contractors, realtors, and merchants," according to Steven A. Riess.[99] These included *Los Angeles Times* publisher and real-estate tycoon Harry Chandler.

The CDA, acting with a dogged and cheerful optimism characteristic of the American booster in a coming-of-age city, secured the 1932 Olympics for Los Angeles. This was a startling upset: Los Angeles was "only the tenth largest city and hardly qualified as a cosmopolitan international locale," as Riess writes.[100] But the Los Angelenos had a winning self-confidence, and they had told the Olympic Committee that a grand outdoor venue, seating 75,000, would be constructed to host the main events.

This arena had been under discussion for several years. Los Angeles was a major league city, said its advocates, with minor league sports facilities. The Angels of the Pacific Coast League played in a ballpark that sat 25,000, while the storied Trojans of the University of Southern California hosted their gridiron tilts in a stadium with a capacity of a measly 7,800.

It is one thing to dream big; it is another to amass the capital with which to make those big dreams reality. The men of the CDA were not unwilling to look to government for support. As Riess notes, "there were very few municipally operated stadiums in America" in 1920, the most prominent, perhaps, being in San Diego. Nevertheless, Southern California was a land of innovation and creativity, and so in August 1920 the voters of Los Angeles had the chance to vote on whether to issue bonds to raise money to build a massive sports arena, an amphitheater, and a municipal auditorium. Consistent with their "reputation for niggardliness" — which might also be considered a solicitous concern for economy in government — they rejected the plan.[101]

But the dream was only deferred briefly. The CDA set forth an alternative plan under which the city and county governments leased acreage at Exposition Park (near USC) to the CDA, which would build a stadium financed by "rents" paid by the city and county. (The rents would amount to half a million dollars over five years for both government entities.) The CDA's influence-wielding members arranged for below-interest construction loans from local banks.

This plan required no public approval via referendum, and in fact the opposition was scattered and powerless. The good-government types at the Municipal League pointed out that the citizenry had already spoken, and in the negative, on the need for a huge sports arena, and it was an affront to democracy for the CDA to act as an unresponsive oligarchy and overturn the expressed sentiment of the people. But the CDA was learning a lesson that many movers and shakers have relearned since: democracy is the enemy of the publicly funded sports stadium, and if you really want to build such a thing you will find a way — whether sly, clever, deceptive, or sledgehammer-like — to step around or stifle public opposition.

In 1923, there opened the Los Angeles Coliseum, "the most expensive sports facility in the country with the exception of the privately financed Yankee Stadium and...the largest sports arena in the United States."[102] Its primary tenant for its first decade was the USC Trojans football team, which according to Riess accounted for 86% of the gate in an average year. USC was a private institution which was benefitting from a publicly subsidized stadium, but times were flush and besides, many pillars of the Los Angeles establishment were USC alumni.

After further public appropriations, including monies raised by a bond issue approved statewide in 1928 to ready the Coliseum for the 1932 Olympiad, the total cost of the arena was $1.9 million, which was less, as Riess notes, than the contemporaneous stadiums in Cleveland (Municipal Stadium, $2.5 million) and Chicago (Soldier Field) cost. The Olympics were a success, after which control of the Los Angeles Coliseum was transferred to the Los Angeles Board of Playground and Recreation Commissioners. This board, now having such a prestigious venue in its portfolio, forthwith changed its name to the Coliseum Commission. This seemed more dignified, as "Playground" was redolent of sandboxes and teeter-totters.

As for Cleveland, in 1930–1931 the city by the lake built Cleveland Municipal Stadium, a massive park with a capacity of over 80,000, financed by a $2.5 million bond issue approved by voters in the early years of the Great Depression. The hope was to attract the Olympics — it didn't work.

But Cleveland's park did keep the major league baseball Indians from moving. The Indians had been unhappy in a ballpark that had a limited seating capacity. They were overjoyed to move into a huge ballpark at public expense. The park's first baseball game, played on July 31, 1932, pitted the Indians against the Philadelphia Athletics and drew a then-record crowd of 80,184.

But Municipal Stadium, buffeted by cold winds off Lake Erie, was a chilly, cavernous, ugly park which the Indians partially abandoned in 1934 (playing all but Sunday and holiday games at League Park) and didn't return to full-time till after World War II. It was cavernously empty during the many years of bad Indians play. Yet, it held one true distinction. In baseball historian Michael Benson's words, "It was the first municipally owned Major League Baseball stadium — and it set quite a precedent."[103]

As the 1960s dawned, so did a period in which the old system by which ballparks were financed had become a quaint archaism. "No longer is the decision to build a stadium made by a private individual who has his personal wealth rising or falling with the worthiness of the decision," wrote Dean V. Baim.[104] No, progress has lifted this burden from the back of the owner and shifted it to the collective back of the taxpayers. And this brave new world is where we pick up our story.

Notes

1. Michael Benson, *Ballparks of North America: A Comprehensive Historical Encyclopedia of Baseball Grounds, Yards and Stadiums, 1845 to 1988* (Jefferson, NC: McFarland & Company, 1989), p. xxvi.
2. Harold Seymour, *Baseball: The Early Years* (New York: Oxford University Press, 1989/1960), p. 48.
3. Warren Goldstein, *Playing for Keeps: A History of Early Baseball* (Ithaca, NY: Cornell University Press, 1989), p. 114.
4. Quoted in Harry Ellard, "Baseball in Cincinnati: A History," www.baseballchronology.com.
5. Harold Seymour, *Baseball: The Early Years*, p. 52.
6. John Wilson, *Playing by the Rules: Sport, Society, and the State*, p. 126.
7. Harold Seymour, *Baseball: The Early Years*, p. 52.

8. Dennis Purdy, *Kiss 'em Goodbye: An ESPN Treasury of Failed, Forgotten, and Departed Teams* (New York: ESPN Books, 2010), p. 215.
9. Harold Seymour, *Baseball: The Early Years*, p. 192.
10. Steven A. Riess, "The Baseball Magnates and Urban Politics in the Progressive Era: 1895–1920," *Journal of Sport History*, Vol. 1, No. 1 (1974): 41.
11. Ibid.: 43.
12. Michael Benson, *Ballparks of North America: A Comprehensive Historical Encyclopedia of Baseball Grounds, Yards and Stadiums, 1845 to 1988*, p. 13.
13. Steven A. Riess, "Historical Perspectives on Sport and Public Policy," *Review of Policy Research*, Vol. 15, No. 1 (1998): 15.
14. Steven A. Riess, "Professional Sunday Baseball: A Study in Social Reform, 1892–1934," *Maryland Historian*, Vol. IV (Fall 1973): 95.
15. Ibid.: 97.
16. Ibid.: 101–103.
17. Steven A. Riess, *Touching Base: Professional Baseball and American Culture in the Progressive Era* (Westport, CT: Greenwood, 1980), p. 95.
18. Michael Benson, *Ballparks of North America: A Comprehensive Historical Encyclopedia of Baseball Grounds, Yards and Stadiums, 1845 to 1988*, p. 304.
19. Ibid., p. 307.
20. Harold Seymour, *Baseball: The Golden Age* (New York: Oxford University Press, 1989/1971), p. 51.
21. Michael Benson, *Ballparks of North America: A Comprehensive Historical Encyclopedia of Baseball Grounds, Yards and Stadiums, 1845 to 1988*, pp. 312–13.
22. Ibid., pp. 22–23.
23. See David Hagberg, "Baltimore Baseball and Beer," www.baltimorebottleclub.org.
24. "Plan $6,000,000 Park at Coogan's Bluff," *New York Times*, May 5, 1911.
25. John Saccomon, "John Brush," The Baseball Biography Project, http://bioproj.sabr.org.
26. "Andrew Freedman," BaseballLibrary.com.
27. Michael Benson, *Ballparks of North America: A Comprehensive Historical Encyclopedia of Baseball Grounds, Yards and Stadiums, 1845 to 1988*, p. 264.
28. Steven A. Riess, "The Baseball Magnates and Urban Politics in the Progressive Era: 1895–1920": 54.
29. Michael Benson, *Ballparks of North America: A Comprehensive Historical Encyclopedia of Baseball Grounds, Yards and Stadiums, 1845 to 1988*, p. 265.
30. Steven A. Riess, *Touching Base: Professional Baseball and American Culture in the Progressive Era*, p. 72.
31. Nelson W. Aldrich, *Old Money: The Mythology of Wealth in America* (New York: Allworth Press, 1997), p. 246.
32. John Harper, "Yankee Stadium Opens," www.nydailynews.com.
33. Harvey Frommer, "Remembering Yankee Stadium: Opening Day 1923," http://baseballguru.com.
34. "Yanks Pick Firm to Build Stadium," *New York Times*, April 19, 1922.
35. Michael Benson, *Ballparks of North America: A Comprehensive Historical Encyclopedia of Baseball Grounds, Yards and Stadiums, 1845 to 1988*, p. 63.

36. Lee Elihu Lowenfish, "A Tale of Many Cities: The Westward Expansion of Major League Baseball in the 1950's," *Journal of the West*, Vol. 17, No. 3 (July 1978): 77.
37. Neil J. Sullivan, *The Dodgers Move West* (New York: Oxford University Press, 1987), pp. 39–40.
38. Ibid., p. 44.
39. Ibid., p. 49.
40. Quoted in ibid., p. 56.
41. Ibid., p. 80.
42. Thomas S. Hines, "Housing, Baseball, and Creeping Socialism: The Battle of Chavez Ravine, Los Angeles: 1949–1959," *Journal of Urban History*, Vol. 8, No. 2 (February 1982): 130–31.
43. Ibid.: 133, 136.
44. Ibid.: 134.
45. Neil J. Sullivan, *The Dodgers Move West*, p. 88.
46. Ibid., p. 85.
47. George Lipsitz, "Sports Stadia and Urban Development: A Tale of Three Cities," *Journal of Sport and Social Issues*, No. 8, Vol. I (1984): 9.
48. Thomas S. Hines, "Housing, Baseball, and Creeping Socialism: The Battle of Chavez Ravine, Los Angeles: 1949–1959": 149.
49. Cary S. Henderson, "Los Angeles and the Dodger War, 1957–1962," *Southern California Quarterly*, Vol. 62, No. 3 (Fall 1980): 268.
50. Neil J. Sullivan, *The Dodgers Move West*, p. 117.
51. Ibid., p. 68.
52. Arthur T. Johnson, "Municipal Administration and the Sports Franchise Relocation Issue," *Public Administration Review*: 521.
53. Cary S. Henderson, "Los Angeles and the Dodger War, 1957–1962," *Southern California Quarterly*: 263.
54. Barry Jacobs, "Sentimental Journey: Brooklyn after the Dodgers," *New York Affairs*, Vol. 7, No. 4 (1983): 140.
55. Cary S. Henderson, "Los Angeles and the Dodger War, 1957–1962," *Southern California Quarterly*: 263.
56. Neil J. Sullivan, *The Dodgers Move West*, p. viii.
57. Ibid., p. 123.
58. Ibid., p. 138.
59. Ibid., p. 153.
60. Cary S. Henderson, "Los Angeles and the Dodger War, 1957–1962," *Southern California Quarterly*: 282.
61. Ibid.: 282–84.
62. Neil J. Sullivan, *The Dodgers Move West*, p. 195.
63. Robert A. Baade, "Stadiums, Professional Sports, and Economic Development: Assessing the Reality," Heartland Institute of Chicago Policy Study No. 62, April 4, 1994, p. 24.
64. Barry Jacobs, "Sentimental Journey: Brooklyn after the Dodgers," *New York Affairs*: 139.
65. Ibid.: 141.

66. Ibid.: 142.
67. Neil J. Sullivan, *The Dodgers Move West,* p. x.
68. Robert F. Bluthardt, "Fenway Park and the Golden Age of the Baseball Park, 1909–1915," *Journal of Popular Culture,* Vol. 21, No. 1 (Summer 1987): 50.
69. Ibid.: 43.
70. Ibid.: 51.
71. Ibid.: 45.
72. Michael Benson, *Ballparks of North America: A Comprehensive Historical Encyclopedia of Baseball Grounds, Yards and Stadiums, 1845 to 1988,* p. 408.
73. George C. Rable, "Patriotism, Platitudes and Politics: Baseball and the American Presidency," *Presidential Studies Quarterly,* Vol. 19, No. 2 (Spring 1989): 363.
74. Michael Benson, *Ballparks of North America: A Comprehensive Historical Encyclopedia of Baseball Grounds, Yards and Stadiums, 1845 to 1988,* p. 410.
75. Bill Veeck with Ed Linn, *Veeck as in Wreck: The Autobiography of Bill Veeck* (Chicago: University of Chicago Press, 2001/1962), p. 285. The puckish Veeck enjoyed roasting his fellow owners' hypocrisy. "We play the Star Spangled Banner before every game," he cracked. "You want us to pay taxes too?" Quoted in Dennis Coates and Brad R. Humphreys, "The Stadium Gambit and Local Economic Development," *Regulation,* Vol. 23, No. 2 (June 2000): 15.
76. James Edward Miller, "The Dowager of 33rd Street: Memorial Stadium and the Politics of Big-Time Sports in Maryland, 1954–1991," *Maryland Historical Magazine,* Vol. 87, No. 2 (Summer 1992): 190–91.
77. Bill Veeck with Ed Linn, *Veeck as in Wreck: The Autobiography of Bill Veeck,* p. 288.
78. Lee Elihu Lowenfish, "A Tale of Many Cities: The Westward Expansion of Major League Baseball in the 1950's": 72.
79. Ibid.: 73.
80. In the 1950s, the moves of Boston to Milwaukee, New York to San Francisco, and Brooklyn to Los Angeles were approved by unanimous votes, as was the move of St. Louis to Baltimore after Bill Veeck sold the team. Veeck's earlier attempts to move were twice rejected. Philadelphia's move to Kansas City was approved by 7–1. James Quirk, "An Economic Analysis of Team Movements in Professional Sports," *Law & Contemporary Problems,* Vol. 38, No. 1 (Winter-Spring 1973): 50.
81. Ibid.: 47, 52.
82. Dean V. Baim *The Sports Stadium as a Municipal Investment* (Westport, CT: Greenwood Press, 1994), p. 21.
83. Lee Elihu Lowenfish, "A Tale of Many Cities: The Westward Expansion of Major League Baseball in the 1950's": 74.
84. Ibid.: 75.
85. Glen Gendzel, "Competitive Boosterism: How Milwaukee Lost the Braves," *Business History Review,* Vol. 69, No. 4 (Winter 1995): 536–37.
86. Ibid.: 540.
87. Ibid.: 549.
88. Douglas S. Powell, "Is Big League Baseball Good Municipal Business?" *The American City* (November 1957): 111.
89. Ibid.: 112.

90. Ibid.: 113.
91. Neil J. Sullivan, "Major League Baseball and American Cities: A Strategy for Playing the Stadium Game," in *The Economics and Politics of Sports Facilities,* edited by Wilbur C. Rich (Westport, CT: Quorum Books, 2000), p. 175.
92. Harold Seymour, *Baseball: The Golden Age,* p. 402.
93. Harold Seymour, *Baseball: The People's Game* (New York: Oxford University Press, 1990), p. 70.
94. Michael Benson, *Ballparks of North America: A Comprehensive Historical Encyclopedia of Baseball Grounds, Yards and Stadiums, 1845 to 1988,* p. 17.
95. "Milestones," *Time,* May 26, 1947.
96. Steven A. Riess, "Historical Perspectives on Sports and Public Policy," in Rich, *The Economics and Politics of Sports Facilities,* p. 18.
97. Perry Duis and Glen E. Holt, "The Classic Problem of Solider Field," *Chicago* (April 1978): 170.
98. Ibid.: 172–73.
99. Steven A. Riess, "Power Without Authority: Los Angeles' Elites and the Construction of the Coliseum," *Journal of Sport History,* Vol. 8, No. 1 (Spring 1981): 51–52.
100. Ibid.: 53.
101. Ibid.: 54–55.
102. Ibid.: 56.
103. Michael Benson, *Ballparks of North America: A Comprehensive Historical Encyclopedia of Baseball Grounds, Yards and Stadiums, 1845 to 1988,* p. 110.
104. Dean V. Baim *The Sports Stadium as a Municipal Investment* (Westport, CT: Greenwood Press, 1994), p. 1.

Chapter 4

Parks and Stadiums Since 1960

The 1960s began in baseball with a New Frontier of its own: expansion franchises were awarded in 1961 to Los Angeles (the Angels) and Washington, DC (a new Senators team to replace the Minnesota-departed one) and in 1962 to New York (the Mets) and Houston (the Colt .45s, later the Astros). So there were four new teams, but only one was placed in a major-league-less city.

The first of the mobile franchises of the previous decade, the Milwaukee (nee Boston) Braves, went south in 1965 to Atlanta, the first city of the Deep South to get a major league team. Indeed, the closest cities to Atlanta in the majors were Houston, Cincinnati, and Baltimore — hardly easy car rides.

Historian Glen Gendzel, analyzing the move in *Business History Review*, chalks the Braves' relocation up to the "competitive boosterism" of Atlanta.[1]

In losing the Braves, Gendzel notes of Milwaukee, "For the first time in modern history, a city was stripped altogether of its major league status."[2] Not for long, though — Milwaukee was to steal the Seattle Pilots out of bankruptcy court in 1970, luring the Pilots into Wisconsin airspace to play at lackluster old Milwaukee County Stadium. Apparently it is one thing to lose a team to a parvenu city that offers lots of goodies, but it is quite another thing to lure a team from another city with lots of goodies of your own. Or so baseball ethicists seem to believe.

The Chicagoans who had bought the Milwaukee Braves from Lou Perini found the offer by Atlanta, "the city too busy to hate," too rich to refuse. And they had sure made a mess of things in Milwaukee. In October 1964,

J.T. Bennett, *They Play, You Pay: Why Taxpayers Build Ballparks, Stadiums, and Arenas for Billionaire Owners and Millionaire Players*, DOI 10.1007/978-1-4614-3332-3_4,

co-owner William C. Bartholomay told the *Milwaukee Sentinel*, "We are positively not moving." In fact, Bartholomay and his ownership group had already agreed to move the team to Atlanta. When they made the announcement weeks later, local reaction was predictably fierce. As Gendzel notes, a Milwaukee third-grader sent Bartholomay a note with YOU ARE A LIAR scrawled in crayon.[3]

Unfortunately for the Braves' owners, the team had a lease with Milwaukee County Stadium that ran through 1965. The county refused to be bought out of that lease, forcing the Braves to play a dreadful lame-duck 1965 season in which attendance plummeted to 555,584. Given the generous lease and the lack of fans to purchase concessions, the county lost money in the lame-duck year, but it was worth it to vengeful Milwaukeeans: the Braves owners lost $1 million (back when that was real money).

Milwaukee, meanwhile, lost a team and, in the minds of its business leaders, its status as a major-league city. "All those krauts want to do is sit around and drink beer": this would be the national image of a Braves-less Milwaukee, or so a *Milwaukee Journal* reporter complained. Hatred of Atlanta was at white heat. As Glen Gendzel remarks, "sullen Milwaukee boosters considered erecting a statue honoring General Sherman."[4]

Baseball Goes South

When the Milwaukee Braves hightailed it to Atlanta, it was to a commerce-minded Southern city that had been desperate for a big league team for years. Atlanta had tried to entice Charley Finley's Kansas City A's to town in the early 1960s, but Finley, like Bill Veeck before him, was unpopular with his fellow owners, who would not permit him to move his team — yet. Atlanta had also been in line for a franchise in the American Football League, but lack of a suitable field led the AFL to redirect that franchise to Oakland. Mayor Ivan Allen Jr. explained the city's position: "We had to find a club which might want to move, then persuade them to move to Atlanta, where we offered them a stadium not yet designed, to be built with money we didn't yet have, on land we didn't yet own." The combined forces of City Hall, Coca-Cola, and the Chamber of Commerce lobbied the Board of Aldermen hard — and they got their stadium.

Using pre-fab construction techniques, Atlanta-Fulton County Stadium was assembled in a year (1964–1965) for $18 million, all of it from public sources. It was "a rush job, and architecturally bland," writes Michael Benson, but it would seat over 53,000 spectators and host the Atlanta Braves and the NFL's expansion Atlanta Falcons beginning in 1966.[5] (The baseball team was supposed to move into its new digs in 1965, but legal wrangling kept the Braves in Milwaukee for a lame-duck year. The new stadium played host in 1965 to an International League AAA team with the decidedly politically incorrect name of the Atlanta Crackers.)

The stadium, one of the early breed of multipurpose cookie-cutters, was criticized for a subpar playing field and the excessive distance of many seats from the action. Moreover, historian of the South David Goldfield remarks, Atlanta-Fulton County Stadium was built "at public expense while the city's abundant poor required special appeals to secure what was left over."[6]

The Braves were greeted as welcome evidence that Atlanta had made the big time. Led by Hammerin' Hank Aaron, the soft-spoken outfielder who would break Babe Ruth's career home run record, the Braves enjoyed enthusiastic fan support. Until the franchise fell into the pits in the mid-1970s. The teams were awful and the stadium, which had all the charm of a quickly built government project, was empty. On one infamous night in 1976, only 970 fans braved tedium to watch a Braves-Houston Astros doubleheader. From 1965 until 1990, "the Atlanta Braves compiled the worst 25-year record in the history of U.S. professional sports."[7]

The buccaneer capitalist Ted Turner bought the team in 1976. At first he tried to lure fans with the wackiest promotions since the heyday of Bill Veeck: quarter-beer nights, Wedlock and Headlock Night blending marriage ceremonies and pro wrestling, and ostrich races. Turner even managed a game himself before the humorless commissioner put an end to that. But there was a method to Turner's madness. He was also assembling an outstanding team under a brilliant coach, Bobby Cox. By the 1990s, the Braves were the most consistently good team in Major League Baseball. They were too good for Atlanta-Fulton County Stadium, which the Falcons of the NFL had left in 1991. It's not like they were leaving a goldmine: the stadium lost $20 million over its first 25 years.

So it was big business and government to the rescue. Atlanta hosted the 1996 Summer Olympics, a public relations disaster which cost taxpayers about $500 million and left the world with the impression that Atlanta had no personality other than a blandly obnoxious craving to be considered "big time." But it left Atlanta with a much better baseball stadium. NBC, which carried the games on television, and other Olympic sponsors footed $170 million of the $209 million bill to construct Centennial Olympic Stadium, the primary outdoor venue of the 1996 Games. After the games, Centennial Olympic Stadium was converted into the baseball-only Turner Field. Across a parking lot from Turner Field stood poor unlamented Atlanta Fulton County Stadium, its days numbered. It was demolished in 1997 over the protests of a Save Our Stadium citizens group.

But the much-traveled Braves franchise had found a home, thanks to an Olympics that is generally regarded as one of the worst of modern times.

The Other LA Team

Just south of the Dodgers, nestled in the antitax hotbed of Orange County, the California (nee Los Angeles) Angels, who had begun life in 1961 in the cozy cradle of LA's Wrigley Field and then shared the Chavez Ravine site with the Dodgers, moved into their home in 1966. Sharing a facility, while it may make a certain economic sense, condemns the less successful team to play in the shadows of the better team, and if the city or state treasury can be raided for construction funds, why not build a brand new stadium, a field of one's own?

The city of Long Beach tried to entice owner Gene Autry and his Angels, but their offer of a city-built stadium included the proviso that "Long Beach" must be part of the team's name. The Long Beach Angels? Uh, no thank you, said Autry. Sounded bush league. The city of Anaheim, by contrast, was perfectly willing to put up with the indignity of hosting a team that was ashamed to identify its host city by name. Assisted by Anaheim's Walt Disney, who sat on the team's advisory board, the city's leaders negotiated the removal of the California Angels to a stadium to be built in Anaheim and paid for entirely by the government. Even hyper-Republican Orange County, the West Coast redoubt of Goldwater conservatism, had bought into the belief that sports facilities were too grand an enterprise to be left to the meager resources of good old free enterprise.

Anaheim Stadium, which opened in 1966, was built on farmland for the Angels by the city of Anaheim, using the construction firm of Del Webb, at a cost of $24 million. It was not a terribly charming park, though as Michael Benson notes, it was — aptly for Southern California — "one of the easiest ballparks to get into and out of by car," surrounded as it was by a web of freeways and highways and entrance and exit lanes.[8] Twenty thousand seats were added in 1979–1980 in order to make the stadium a viable home for the Los Angeles Rams of the NFL. When the Rams took off for St. Louis, the Angels demanded that the stadium, eventually to be known by the rather pretentious name of Angels Stadium of Anaheim, be reconfigured into a baseball-only park with club-level and dugout-level suites, a modernized press box, and three full-service restaurants. The cost of these 1997–1999 improvements: $117 million. The Disney Corporation, which had a controlling interest in the Angels, footed over three-quarters of the bill ($87 million), while the remaining $30 million was billed to the taxpayers of Anaheim, most of it supplied through hotel taxes and the city's reserve funds.

Were the Angels grateful to Anaheim? If so, they chose a strange way of showing it. Since its inception, the franchise had gone through names like Elizabeth Taylor used to go through husbands. Born in 1961 as the Los Angeles Angels, they became the California Angels in 1965 and then the Anaheim Angels in 1997. In 2005, owner Arte Moreno, who had bought the team from Disney in 2003 for $180 million, changed the name to the verbose "Los Angeles Angels of Anaheim." This was done to meet the requirements of the team's lease that the ballclub's name must include the name of the city in whose confines it plays. But while the team's home city (and benefactor) remained part of the name, it was as the caboose. From now on, sports pages and sports announcers would refer to them as the "Los Angeles Angels," the American League's saintlier — or deader? — counterparts of the National League's Los Angeles Dodgers. Thanks for the subsidy, Anaheim! A lawsuit by the city against the Angels failed, and now no one outside of the Anaheim Chamber of Commerce ever uses the A word when placing the Angels. The moral: draw your lease tighter. Or don't fork out the subsidy in the first place.

It is an open question whether the city of Anaheim has learned that lesson. The ever-dangling prospect of bringing an NFL expansion (or relocated) team to Los Angeles has drawn some interest from the city that was good enough for Disney but not for Arte Moreno, or at least not for Arte to utter the name. The parking lot of Angel Stadium has been proposed as a prime site for a football stadium in what giddy city planners call "an urban environment of a scale never before seen in Orange County."[9]

Huge government-constructed stadiums are not everyone's idea of an "urban environment," but in any case Anaheim was not without skeptics. Among those weighing in when the city started considering the possibility of a football arena in 2005 was University of Chicago economist Allen Sanderson, who laid it out for the *Los Angeles Times* with all the force of a Ronnie Lott hit on a wide receiver going over the middle: "There are only two things you do not want on a valuable piece of real estate. One is a cemetery, and the other is a football stadium." As Sanderson pointed out, the football stadium lies dormant for almost as many days — at least 353, depending upon a team's playoff run — as does the cemetery. The mayor of another potential site, Roosevelt Dorn of Inglewood, said wisely, "The last thing we would want is a football stadium in Inglewood. It's a white elephant."[10]

Would that more mayors could spot the white elephants in the jungle of boosterism.

Meet the Mets

The New York Metropolitans, or Mets, began life in 1962 as the lovable losers of the National League, posting the worst record of modern times, 40–120. The city of New York built William A. Shea Municipal Stadium in the Flushing neighborhood of Queens to house the team in a low-rent deal. The Mets fumbled and whiffed their way through the 1962 and 1963 seasons at the Polo Grounds, but by April 1964 Shea Stadium, named for the man who brought National League baseball back to New York — and in the same section of town as Robert Moses had wanted the Brooklyn Dodgers to relocate — opened its doors to a curious city. The city expended $28.5 million to build Shea for an owner, Mrs. Joan Whitney Payson, who was an heiress descended from the wealthy and aristocratic Whitney and Hay families, and a collector of the artwork of Renoir, Cezanne, Monet, and many other impressionist painters. Just why the entertainment business owned by such a woman should be deserving of such an enormous public subsidy the solons of New York never quite explained.

While Shea Stadium was generally regarded as a charmless dump, the Mets electrified the city of New York and the country as a whole with their 1969 pennant and World Series championship. Shea was also home to the New York Jets, who enjoyed a brief vogue when man about town Joe Namath quarterbacked the team in the late 1960s. But publicly built Shea Stadium never quite achieved the hold on the Gothamite imagination that privately built Yankee Stadium had, and when in late 2008 Shea was demolished, the river of tears over its implosion probably couldn't have filled a Dixie cup. The site of Shea is now a parking lot for the Mets' Citi Field, which the taxpayers of New York paid for — grudgingly, in many cases — through the city's sale of municipal bonds, which are to be paid off by the Mets.

While the Mets project came in at about $830 million — a bargain compared to the simultaneous boondoggle in the Bronx for the Yankees — the various rebates and exemptions and lease breaks and outright subsidies reduced the amount put up by the Mets ownership (led by Bernie Madoff-investor Fred Wilpon) to only $135 million, in Neil deMause's estimation.[11] Citi Field opened in 2009, the same year as the new Yankee Stadium hosted its first game. Citi Group, which pays $20 million a year to keep its name on the stadium, was a beneficiary of the 2008 Troubled Asset Relief Program (TARP) bailout, leading some wags to suggest that the park be renamed Citi/Taxpayer Field. But Major League Baseball is big business, and big business — especially big business that is, essentially, a subsidiary of big government — tends to lack a sense of humor. The name stayed.

Washington Goes National

In 1962, the reborn Washington Senators, who had spent their first year in the privately built old Griffith Stadium, moved into District of Columbia Stadium, later renamed Robert F. Kennedy (RFK) Stadium after the slain senator and brother of the late president. RFK, which would at various times be home to the chronically cellar-dwelling Senators and the football Redskins, cost $18.756 million to build. The construction bonds were paid equally by the federal government and the government of the District of Columbia.

The Senators skulked out of town after the 1971 season, becoming the Texas Rangers. And when Major League Baseball revisited Washington yet again in 2005, the team name was depoliticized from Senators to Nationals, but the method of funding owed more to RFK Stadium than to Griffith Stadium.

The franchise that became the Nationals was born in 1969 as the Montreal Expos. The Expos came into the league as the great Canadian hope, the first major league team north of the border and the property of Charles Bronfman of Seagram's. Montreal had a long history in the AAA International League: Jackie Robinson had played here in 1946 for the Montreal Royals, a Brooklyn Dodgers farm team. Playing in cozy if oft-derided Jarry Park, the Expos were the toast of the town — though a decidedly secondary toast to the Montreal Canadians of the National Hockey League. The Expos moved into Olympic Stadium, which was built as the main venue for the 1976 Summer Olympics, and played in its confines until Major League Baseball left town in 2004.

Though plagued in later years by poor attendance, the Expos were also one of those franchises forever griping about their stadium. Not that they built it on their loonie or toonie. Olympic Stadium was a project of the province of Quebec. Built between 1973 and 1976 for what the government estimated would be a cost of $264 million (Canadian), it was not paid off until 2006, when the final bill came in at $1.61 billion. Olympic Stadium was a boring white elephant whose nicknames have included The Big-Oh and The Big Mistake. It was not a mistake Quebec would make twice.

An art dealer named Jeffrey Loria became the controlling interest in the Expos in 1999; he requested that the province of Quebec build the Expos a new stadium to replace the graceless Big-Oh, but the ruling Parti Quebecois said *non*. Montreal was not about to cave in to the Expos' demands. Prime Minister Jean Chretien said, "We're not in the business of helping sports teams."[12]

In 2002, Major League Baseball purchased the Expos from Loria, hoping to euthanize the team and contract the league — a vindication of union leader Donald Fehr's opinion that the league wants to keep a ready supply of team-less cities out there bidding for franchises on the public dime. The league did not contract. Instead, it dangled the prospect of the big leagues before the salivating city fathers of Portland, Charlotte, San Juan, Norfolk (VA), and the DC metro area before issuing from on high the pronouncement that in 2005 the Expos would metamorphose into the Washington Nationals, who would play their first three seasons at RFK.

While Washington-area baseball fans were thrilled at the news that the majors were returning — for one thing, they had tired of having to cheer for the hapless Baltimore Orioles up I-95 — their excitement was not so outrageously contagious as to lead to broad support for supplying the city's newest resident with a brand new home. A *Washington Post* poll in November 2004, in the fall before the Expos came to town, found that "more than two-thirds of District residents [69 percent] oppose using public funds to build a baseball stadium in the city." The *Post* characterized the opposition as "both broad and deep."

Washington Mayor Anthony Williams was the most prominent supporter of a taxpayer-subsidized stadium for the Nationals. (As 57-year-old bus driver Ceasar Short told the *Post*, Williams "wanted baseball too much. They knew this guy, he was a pushover when it came to baseball.")[13] Mayor Williams, on October 1, 2004, issued a message in which he proposed a financing plan in which the onus of paying for the Nats' new park would fall on "team owners, those who use the ballpark, and...D.C.'s largest businesses."[14] As it turned out, and as one would expect, the "contribution" of business via a new tax district dwarfed that of the owners.

The elite of the city were largely united in support of constructing a new stadium to replace RKF. Opposition came largely from powerless grassroots groups such as Black Voices for Peace, whose leader, Damu Smith, said, "We say to this member and some members of the city council that you will not get this stadium when people are sleeping on the streets and schools are crumbling." Mayor Williams dismissed complaints such as this as "populist grandstanding."[15]

The City Council was somewhat less enthusiastic than was Mayor Williams, at one point voting to limit the city's contribution to any new stadium to half the construction cost. Tough talk, but it was not backed by tough action.

The Lerner family, whose patriarch, real estate developer Ted Lerner, had been named by *Forbes* magazine one of the 400 richest men in America (net wealth: $3 billion), purchased the Nationals from Major League Baseball for $450 million in 2006. The publicity-shy Ted Lerner comes off as a martinet in a 2006 *Washington Post* profile of him. No one will say a cross word about him, exactly, for they know that there will be a price to pay. However, a deceased former business partner, H. Max Ammerman, said of Lerner shortly before Ammerman died, "If he didn't get his way, he was a terror. Meetings would end up in verbal fights." Lerner once fired his own brother, who claimed that he had been "cheated out of tens of millions of dollars over the years."[16]

A delightful gentleman, no doubt, and the city of Washington should not have been surprised that Lerner drove a hard bargain in seeking the best possible terms for his team. The City Council's attempt to keep the public contribution to the stadium below 50% was about as weak and futile an effort as a Little Leaguer might give in swinging at one of Nationals' star pitcher Stephen Strasburg's fastballs. That 50% eventually rose to almost 100%.

Nationals Park, which cost $611 million to construct, was opened in the spring of 2008. Owned by the District of Columbia Sports Commission, the park is financed by city bonds. The largest single revenue source is — surprise! — a gross receipts tax on new businesses in the District. Second largest source is from in-stadium taxes on tickets, concessions, and merchandise.

The ballpark deal is widely regarded as a one-sided joke. As D.C. Councilman-at-Large David A. Catania said, "What I saw was no negotiations from the mayor. He basically threw open the gates of the treasury and said, 'Help yourself.'"[17] No less than Marion Barry, the scamp of a former mayor, called the DC baseball deal "the biggest stick-up since Jesse James and the great train robbery."[18]

The Nationals took the city for everything they could, and in return Washingtonians got a last-place ball team and an admittedly attractive park. City residents can't say they weren't warned. In October 2004, as Mayor Williams was campaigning for his stadium deal, 90 economists signed and published an open letter to the mayor and city council "on the likely impact of a taxpayer-financed baseball stadium in the District of Columbia."

"A vast body of economic research," began the economists, "suggests that the proposed $440 million [costs, inevitably, rose] baseball stadium in the District of Columbia will not generate notable economic or fiscal benefits for the city." They restate what has become a truism among economists: that "sports stadiums do not increase overall entertainment spending but merely shift it from other entertainment venues to the stadium."

In the specific case of the Washington Nationals and their proposed palace for the summer game, the stadium "cannot be expected to generate a net increase in economic activity in the Washington metropolitan area." It may very well "shift some entertainment spending from the Maryland and Virginia suburbs into the District," but it "alone will not revitalize the Anacostia waterfront."[19] (As has become the fashion, the stadium was sold as a means of imparting new life into a neglected neighborhood; thus, it could be seen not only as a benefit to sports fans but also as a boon to the local economy. Mayor Williams went so far as to say at the groundbreaking ceremony, "This ballpark really is about...the rebirth of the Anacostia waterfront.")[20]

The economists went unheeded. Economists usually do. After an absence of over three decades, Major League Baseball was sniffing around our nation's capital again, and the capital's mandarins could barely hide their excitement. Build a $611 million stadium for a *Forbes* 400 real-estate developer? Why not? America's game belonged in America's city.

One of the economists who signed the open letter, Dennis Coates of the University of Maryland-Baltimore, joined with Brad R. Humphreys of the University of Illinois to challenge the case for DC baseball in a Cato Institute briefing paper of October 2004.

In keeping with what has become an unhallowed tradition of absurd overestimates of a sports venue's impact on the city and its workforce, the DC Office of the Deputy Mayor for Planning and Economic Development asserted that the park would "create 360 jobs earning an annual total of $94 million," which, as Coates and Humphreys note, "amounts to an astounding

$261,111 per job."[21] Even the inflated payroll of a major league team couldn't quite meet that marker. And while the vast majority of the payroll is consumed by the twenty-some members of the Nationals team, very few actually live in the District. Thus, most of the Nats' payroll is not captured by the full force of the DC income tax. (For which the Nats' players are no doubt grateful.)

Coates and Humphreys apply the insights of the last quarter-century's worth of study on the economic impact of pro sports to the DC case. As they note, the alleged new spending by consumers upon the Nats is really a spending transfer. Since the fans are drawn overwhelmingly from DC and its Maryland and Virginia suburbs, what is actually happening is that a fan plunking down $100 for tickets, parking, and a quick meal within or just outside the park is spending money he otherwise would be spending at a Dupont Circle nightclub or an Arlington restaurant. The two economists give this very DC example: "Joan Suburban has a season subscription to the National Symphony Orchestra. Every month or so, she leaves her Silver Spring condo, drives into the District for a few drinks and a meal in a Georgetown restaurant, and enjoys a concert. As she is also a baseball fan, Joan declines to renew her NSO subscription next year and instead buys a partial-season ticket plan for the new team. Every month or so, she leaves her Silver Spring condo, drives into the District for a few drinks and a meal in the revitalized area around the new ballpark, and enjoys a ball game. How much new economic benefit does the District gain?"[22]

While the economic development people, who shoot everything that flies and claim everything that falls, will claim Joan Suburban's spending on Nats' games, the revenue lost by the Georgetown restaurant does not figure into such calculations.

Six years later, the economists appeared prescient. The waterfront is as economically stagnant as ever. The ballpark has almost as many empty seats as paying customers. Average home attendance peaked at 33,728 when the team played at RFK in its inaugural season, 2005. It fell by almost 10,000 over the next 2 years, then edged upward to 29,005 in 2008, when Nationals Park opened. (Capacity is 41,888.) It plunged to 22,716 in 2009, 22,568 in 2010, and 24,877 in 2011 — in the middle year this was despite the excitement surrounding the debut of pitching phenom Stephen Strasburg.

There was, however, one big winner to come out of the deal (besides the Lerner family): the city treasury. As the *Examiner* of Washington reported in June 2010, the gross receipts tax "has become such a cash cow that the city is now using it to help close its nine-figure budget gap." Revenues from the tax are running ahead of the city's bond payments for the stadium to such an extent — a cumulative total of $135 million from 2005 to 2010 — that the city is using the surplus to "plug monstrous holes in the District's budget."

This is contrary to the agreement as understood by those local business leaders who helped shepherd the stadium deal through its political journey. As Barbara Lang, chief executive of the DC Chamber of Commerce, told the *Examiner,* "The deal that we had. . .was that any excess monies would be used to pay down the bond. We would like to see those bonds paid off earlier to relieve us of that tax. I'm very concerned that it will become a part of the city's operating budget."

Her concern is well placed. Nothing is quite so permanent as a "temporary" tax. Once in place, a levy is awfully hard to dislodge. Councilman Jack Evans, who calls the gross receipts tax "the biggest mistake that this government has ever made" — and given that we're talking about the District of Columbia here, *that's saying something* — told the *Examiner* that if the revenue from the tax were applied solely to the stadium debt, as the original deal had it, the bonds would be paid off half as early as the 30-year repayment schedule called for.[23]

And as for Anacostia, "What did [the agreement] bring here besides the stadium?" asked Victor Williams, who lives in the neighborhood.[24] Taxes and losing baseball: you can always count on DC for those.

Gone to Texas

The Washington Senators, meanwhile, went to Texas. The city of Arlington, located between Dallas and Fort Worth, had seemed the ideal spot for a potential major league franchise. So in 1964 Tarrant County spent $1.9 million to build the automobile-culture named Turnpike Stadium, which took its moniker from the Dallas-Fort Worth Turnpike. Built ostensibly for the Dallas Spurs of the Double A Texas League, the park was constructed so as to be easily expandable from 10,000 capacity to five times that amount.

When the perennially cellar-dwelling Washington Senators looked westward for a new dwelling for the 1972 season, their sights settled on Turnpike Stadium, which was quickly expanded to 35,000-plus capacity and redubbed Arlington Stadium. Various modifications to make the park suitable for what were to be called the Texas Rangers were undertaken. The total package offered by Arlington (including infrastructure) was $28.5 million — add to this the interest on the city's bond issue, and Arlington was out $44.3 million for the privilege of hosting the Rangers.

Yet Arlington Stadium was small, it was hot, and it was a pitcher's park. Baseball in the late twentieth century preferred big, preferably domed, hitter's parks. So in 1994 the Rangers moved into the new and grandly named Ballpark in Arlington, located just across the parking lot from Arlington Stadium, which was, without much ceremony, demolished. Poor Arlington Stadium wasn't old, it wasn't hallowed by great games or World Series teams, and it hadn't even had a catchy name. The taxpayers built it, and then they knocked it down.

The Ballpark in Arlington was briefly renamed Ameriquest Field in Arlington before reverting to a modified form of its original name, Rangers Ballpark in Arlington. (Wordy stadium titles are an unfortunate trend of recent years.) It was built for $191 million, toward which the Rangers contributed about 30% of the cost. The other 70% was the responsibility of the taxpayers of Arlington.

The Rangers' general managing partner — the public face and private lobbying voice of the ballclub, even though his 2% ownership stake had come from borrowed money — was George W. Bush, son of the president of the USA. Or at least that was his public description in 1991, when the voters of Arlington were asked to approve a $135 million bond issue to build the park. They did approve it, by a margin of two to one, increasing the sales tax by half a cent. The legislatively created stadium authority that was to oversee this government-owned entity made considerable use of eminent domain to enlarge its own domain. Incredibly, this "rent to own" deal would permit the Rangers eventually to buy the ballpark for a measly $60 million. Baseball writer Andrew Clem writes of the future President Bush's lucrative involvement in this extended episode: "In December 1994 Bush stepped down as managing general partner just prior to being inaugurated governor of Texas, but retained his equity stake in the team. In June 1998 the Rangers were purchased by Tom Hicks for $250 million, the second-highest sum ever paid for a Major League Baseball team. The transaction triggered the contingent 10 percent escalator bonus on top of his two percent equity share, so that Bush received a total of $14.9 million in proceeds for his $606,000 total investment. Not a bad rate of return! Though everything was done above board and within the law, the whole affair does reek of 'stadium socialism' or 'crony capitalism,' as you prefer."[25]

Somehow this seems a fitting sequel for a team once called the Washington Senators.

Tearing Down Ruth's House

Yankee Stadium would be hallowed by eight decades of great players — Ruth, Gehrig, DiMaggio, Mantle, Jeter. The House that Ruth (and Ruppert) built endured through a Great Depression, a Second World War, and the ravages of the government-subsidized cookie-cutter stadiums of the 1970s. It hosted classic fights, such as the Joe Louis–Max Schmeling bout in 1938, as well as New York Giants football, most notably the 1958 overtime NFL title game won by the Giants against the Baltimore Colts, sometimes called — at least by New York chauvinists — "the greatest game ever played."

But Yankee Stadium, which Jacob Ruppert had built entirely with private money, was no longer privately owned. (As historian Neil J. Sullivan writes, in 1923 "no one suggested that city taxpayers should assume the cost of the Yankees' new facility."[26]) In 1953, the then-owners of the team, Dan Topping and Del Webb, sold Yankee Stadium to Chicago financier Arnold Johnson. Johnson in turn sold the stadium to a friend, John Cox, who left it to Rice University in 1962. (The land he had sold earlier to the Knights of Columbus.) So the House that Ruth and Ruppert built was owned, incongruously, by an elite private university in Houston, Texas.

CBS bought the team in 1964, managed it atrociously, and sold it to a syndicate led by George Steinbrenner in 1973. CBS may have been the top-rated television network of the era, but whatever lessons the corporation had learned from selling Walter Cronkite and Buddy Ebsen to the nation's viewers didn't seem to translate onto the baseball diamond. The teams were lousy, and the corporation begged the city of New York to build it a new domed stadium amidst an ocean of parking spaces.

Under Mayor John V. Lindsay, the City of New York bought the stadium from Rice and leased it back to the CBS-owned Yankees in a sweetheart deal. These were the years in which New York was undergoing the fiscal hemorrhage which would lead to the infamous New York *Daily News* headline "Ford to City: Drop Dead," but still, the moribund city underwrote up to $150 million in renovations as Yankee Stadium underwent a major overhaul in the mid-1970s. The Yankees were forced to play at Shea Stadium in Queens in 1974–1975 before moving back into their new and not necessarily improved home. Times had certainly changed. As Sullivan observes, "If no one thought to offer Ruppert public money to build the Stadium, the use of the public purse to pay over $100 million for the renovation was a foregone conclusion to Lindsay and the Yankee owners."[27] Private construction or reconstruction of a Major League Baseball park à la Jacob Ruppert or even Walter O'Malley had become unthinkable, an anachronism on the order of the three-ball walk, the nine-game World Series, and the three-fingered fielder's glove.

Even with the many alterations, this was still Yankee Stadium, for Pete's sake, so when the Yankees and the usual politicos started making noises about demolishing the venerable stadium and replacing it with a palace that Rudy (Giuliani) built, loyal fans and baseball traditionalists raised a protest. All to no avail, for when ambitious politicians have the chance to be publicly identified with the home team — especially a perpetually winning home team — the objections of taxpayers are no object. The fortuitous April 13, 1998, collapse of a 500-pound concrete and steel beam forced the postponement of two games and gave an invaluable boost to those who argued for a new (taxpayer-bought) stadium for the Bronx Bombers. (They were not deterred by the finding of New York City buildings commissioner Gaston Silva, who said, "From a structural perspective, there's no reason why Yankee Stadium can't be around for another seventy-five years if it's maintained properly.")[28]

George Steinbrenner made noises about moving the team to New Jersey or another borough, especially Manhattan, but no one seriously thought he was going to leave the City of New York. He had made these noises for years, such that even the *New York Times,* usually a sucker for any kind of government expenditure, editorialized, "The symbolic impact of losing the team to New Jersey or anywhere else would be incalculable. It would injure New York's prestige and its soul. But fatally? No. The city survived the departure of the Dodgers and Giants, and would surely live on if the Yankees desert."[29]

The Yankees were inextricably *of* New York; based outside the city, they would be just another baseball team. Everyone knew that. But that didn't stop Steinbrenner from demanding subsidies that would easily reach into the hundreds of millions of dollars. As Jim Dwyer wrote in the New York *Daily News,* "If George Steinbrenner wishes to stand up today and say I, like the Brooklyn Dodgers, am willing to use my funds to pay for this thing, then the whole discussion is over. He can build any place he wants. He has no intention of spending his own cash because he has been spoiled rotten for the past 26 years. He is more dependent on government than all the welfare mothers in New York."[30]

George was undeterred. Ground was broken for the new Yankee Stadium on August 16, 2006; the team played its first game on the new field on April 16, 2009. At the groundbreaking, George Steinbrenner said somewhat grandly, and with no evident sense of irony, "It's a pleasure to give this to you people."[31]

While it incorporates many design elements from the old stadium, the twenty-first-century Yankee Stadium deviated from the Colonel Ruppert pattern in its method of financing. No rich beer baron paid for this ballpark. Rather, the cost was split between the Yankees ownership, which promised to contribute $1.1 billion toward the construction of this Taj Mahal of baseball, and the city of New York, which pledged to spend $220 million on parking facilities and infrastructure work on what had been city parkland. In addition, the team enjoys a 40-year exemption from property taxes.

Jim Dwyer, writing in the *New York Times,* captured the evolution — or devolution — of city policy with respect to professional sports: "The first incarnation of Yankee Stadium opened in 1923. The owner, Jacob Ruppert, bought private land, raised private funds for the construction, and maintained the place with money he made in ticket sales... [B]aseball in New York was recognized by the government as another commercial venture, with all the opportunities and responsibilities of owning property."[32] Not any more.

The squeeze on taxpayers could have been tighter had not Mayor Michael Bloomberg drawn in the reins on the plans of his predecessor, Rudy Giuliani. In 2001, his final year in office — despite his unseemly last-minute attempt to repeal mayoral term limits in the aftermath of 9/11 — Mayor Giuliani had promised to build both the Yankees and Mets $800 million retractable roof stadiums, though the teams would have had to pay rent. The well of money was bottomless, it seems, in the Big Apple.

Naturally, the new Yankee Stadium, unlike Colonel Ruppert's cathedral, went vastly over budget, with the jock-worshipping politicians volunteering taxpayers to pick up much of the overrun. Neil deMause, coauthor of *Field of Schemes,* estimates that the Yankees deal ultimately cost the taxpayers of New York City, New York State, and even the United States of America almost *$1.2 billion.* Including upwards of $400 million in additional property tax breaks, this reduced the Yankees total contribution, originally slated to be $1.1 billion, to under $700 million.[33]

New York had come a long way since 1957, when Mayor Robert Wagner responded in this way to the Dodgers' and Giants' threats: "If we began to subsidize baseball teams, all sorts of business enterprises would demand the same things. Our feeling is that professional ball clubs class as private enterprise. They have to carry their own weight. We will not be blackjacked."[34] Mayor Wagner was no Rudy Giuliani, that's for sure.

It's the Pitts

Barney Dreyfuss's enduring monument of Forbes Field endured until 1970, when, as with Philadelphia, Pittsburgh destroyed a privately built baseball palace of the sport's early years and replaced it with a $55 million multipurpose stadium, Three Rivers, which was as drab and generic and unloved as Veterans Stadium was across the state in Philadelphia. And like Veterans — and unlike Shibe Park and Forbes Field — Three Rivers Stadium was a purely public project, an ugly outdoor artificial turf stadium. Also like Veterans, Three Rivers was eventually brought down in a controlled demolition, witnessed by thousands of fans, few of whom could get terribly sentimental over the cookie-cutter park, even though it had been the scene of the glory days of Pittsburgh professional sports, as the Steelers and Pirates had fielded some superb teams throughout the 1970s and 1980s. Like other cookie-cutters, its obsolescence was not so much physical as it was economical and aesthetic.

Three Rivers gave way to Heinz Field (for the NFL Steelers) and PNC Park (for the National League Pirates), which were born not in Barney Dreyfussian dreams of grandeur but in the far more mundane offices of the Allegheny Regional Asset District, the governmental entity which funds libraries, parks, and recreation in Allegheny County, Pennsylvania, whose largest city is Pittsburgh. The RAD stepped in when in November 1997 voters in the 11-county area surrounding Pittsburgh overwhelmingly rejected the carefully named Regional Renaissance Initiative, a plan to "initiate" a "renaissance" in Pittsburgh's stadium scene by jacking up the sales tax to pay for new stadiums for the Pirates and Steelers. The vote was so crushing — the RRI went down by 530,706–281,336 — that it sent architects of the plan back to the drawing board. The lesson they learned from the defeat of RRI: Do *not* put new proposals before the voters. Pittsburgh Mayor Tom Murphy spun the results this way: "You get two messages from the voters: Don't use public money for ball parks to pay for greedy owners, but don't you dare let these teams leave."[35] In other words, even though the voters had overwhelmingly defeated the proposal to build new stadiums for the Steelers and Pirates, these voters were irrational and so their expressed opinions ought to be taken with a large grain of salt. They said no, but they didn't really mean it.

Thus was born Plan B: the Regional Destination Development Plan, an $809 million scheme to build new stadiums for the Pirates and Steelers and to expand the David L. Lawrence Convention Center. Of this $809 million, $305 million was to be raised through existing county sales and hotel taxes, $300 million was to come from state taxpayers, the feds were kicking in $28 million, and the remaining $176 million was to come from private sources, primarily the Pirates and Steelers ownership. The key to passing the Regional Destination Development Plan was to keep the voters as far away from it as possible; this plan did not need voter approval but only ratification by the Regional Asset District, whose members are appointed, not elected. Since all but one member owes his or her appointment to the county executive or the mayor of Pittsburgh, the RAD is unlikely to take the side of the man on the street versus the man who gave its members their appointments. And so on July 9, 1998, by a vote of 6–1, the Allegheny Regional Asset District approved Plan B, and set into motion the construction of the fields whose naming rights would later be purchased by Pittsburgh-based PNC Bank and Heinz, the Pittsburgh ketchup king.

The brave dissenter on the RAD board, Ralph DeStefano, president of a local hospital, denounced the plan as "corporate welfare." He asked, "Why should the elderly and poor be forced to feed this insatiable monster?"

Representing the Good Sports Coalition, the citizens group which had opposed the earlier proposed sales tax increase, Rob Chesnavich asked the RAD board, "What does the public need to do to show we don't want stadiums? Any time the public's been polled on this they've said, 'No.' You're telling the public you don't want their opinion, you want their wallets."

Mr. Chesnavich was exactly right. Putting stadium subsidies up for public vote is fraught with danger for the seekers of corporate welfare. Far better it is to concentrate lobbying on small unelected boards that have, over the years, accumulated extraordinary powers to disburse public monies.

Attorney Allen Brunwasser, another Plan B foe, scoffed that "If anyone came into one of our banks with the kind of arithmetic I've heard tonight, they would be laughed out of the place. If this was a good deal, they wouldn't have to come here to beg for money."[36] Again, the remark was on target and yet fruitless, for agencies such as the RAD exist in order to dole out monies to interests that cannot obtain them through the private market.

The Pirates and Steelers play on in their new fields, greened by taxpayer dollars.

The Orioles Find a Nest

In April 1992, Oriole Park at Camden Yards opened in Baltimore. This was a new baseball park that felt old, praised by architecture critics if bemoaned by taxpayers, who were stuck with 96% of the $110 million bill for the three-year project. Financing was primarily through revenue bonds and lottery proceeds.

As James Edward Miller explains in his study of the business of baseball in Baltimore, efforts in the 1970s to coax or coerce taxpayers into building a new stadium for the Orioles and Colts in the Camden Yards section of Baltimore came to nothing, since the Maryland state legislature refused to kick in state monies. City voters even passed a referendum barring the city from contributing taxpayer monies to new stadium construction. And so the city kept the two franchises in place with piecemeal improvements to Memorial Stadium and "extremely favorable" lease terms. By 1977, the city was losing over $730,000 annually on Memorial Stadium.[37]

Indirectly, Baltimore can blame or credit its resented neighbor to the south, Washington, DC, for the eventual construction of Camden Yards. In 1978, the controversial attorney and political fixer Edward Bennett Williams, a Washington establishment figure, bought the Orioles. He made the usual pledges to keep the team in Baltimore, but he hinted broadly that his goodwill and loyalty depended upon Baltimoreans opening up the city and state coffers. The implied threat was that the Orioles would move to Washington to take the place of the departed Senators, and Baltimore would be without a Major League Baseball team. When Robert Irsay and the Mayflower moving vans carted off the football Colts over the dark evening of March 28–29, 1984, the prospect of losing the Orioles, too, shook loose the Maryland money tree. Led by the bulldoggish Mayor, later governor, Donald Schaefer, Baltimore's "political and business establishment" moved into high gear behind the proposed Camden Yards ballpark.[38] Ground was broken on June 28, 1989, and the Orioles moved in on April 6, 1992. Cost to taxpayers? A cool hundred million. The benefit to owner Edward Bennett Williams? He had bought the team in 1978 for $12 million, and when he died a decade later, with Camden Yards on the horizon, it was sold for $70 million to Eli Jacobs in 1989. Jacobs sold it in 1993 for $173 million. A nice profit, courtesy, in part, of the taxpayers of Maryland who paid for Camden Yards.

As Charles Euchner writes in his detailed survey of the fight over building Camden Yards, the New Deal-style wheeler-dealer Donald Schaefer, who was elected governor of Maryland in 1986, prevailed over an opposition that lacked his (taxpayer-funded) resources — notably, his ability to throw money to Maryland legislative districts outside the Baltimore area. Schaefer also had a compliant judiciary and state attorney general. Marylanders for Sports Sanity (MASS) collected 28,000 signatures demanding a referendum on the Maryland legislature's plan for publicly subsidized baseball and football stadiums, but the Maryland Court of Appeals ruled that these expenditures were not subject to referenda oversight.

"A referendum campaign," writes Euchner, "could have created an open-ended debate about the role of professional sports in the city, subsidies to private interests, the role of neighborhoods in local politics, and other legislative and fiscal priorities for the state and city. The state's opposition to the referendum started, in essence, a discussion about whether full public discussion ought to be permitted."[39] When you have to discuss whether or not you can have a discussion, you're in trouble.

No one — or maybe only those who venerate the Seattle Kingdome and Veterans Stadium in Philadelphia as the apexes of sports architecture — has a bad word to say about Camden Yards, with its comfortably retro design and its views of the reborn Inner Harbor of Baltimore.

Camden Yards, jewel of the new wave of retro ballparks, is sometimes held out as the "exception to the conventional wisdom that publicly financed sports stadiums are bad deals for cities," write Bruce W. Hamilton and Peter Kahn of the Department of Economics at Johns Hopkins University, yet Camden Yards is a drain on Baltimore.[40] Hamilton and Kahn found that in 1997, the Orioles park generated about $3 million in economic benefits while it cost the taxpayers about $14 million annually, for a yearly loss of about $11 million. And that was during the Camden Yards honeymoon. In 1992, during the park's first year, average attendance was 44,598, as opposed to 32,313 in 1991, and the numbers stayed steady through 1996, the last year before their study. Average attendance in recent years has plunged; in 2010 it was only 21,662. Moreover, 70% of the incremental increase in attendance since Camden Yards opened came from fans outside of Maryland (mostly in DC and Virginia), a far higher percentage than other ballparks experience.[41] The park is, in that respect, most atypical.

Camden Yards was envisioned as a key piece of downtown Baltimore's revitalization. Aesthetically pleasing, located downtown rather than in the midst of a bay of suburban parking spaces, Camden Yards was refreshingly new — yet it was also a throwback to the era of Big Government sports facilities, for the "land acquisition and stadium construction costs" were "almost entirely borne by taxpayers," as economists Dennis Coates and Brad R. Humphreys have written.[42] Moreover, Roger Noll estimates that the cost per job created by public funding of Camden Yards was $125,000, which is consistent with the claim by Allen Sanderson of the University of Chicago that "cities would be better off if the mayor were to go up in a helicopter and dump out $100,000" rather than rely on government-subsidized stadiums to bring prosperity.[43]

It's too late to drop money out of a helicopter over Baltimore. The Orioles do have a nice stadium, though. And at least they, unlike the Colts, stuck around...

Indiana-Some-Place

By 1972 the bloom was off the rose for Baltimore Colts' owner Carroll Rosenbloom, who traded his team for the Los Angeles Rams. The Colts now belonged to Robert Irsay, but if any Baltimore fans thought they were in for a more placid era, they were badly mistaken.

Irsay first met with Indianapolis officials in 1977. Seven years later, he and his Colts skipped town.

On March 29, 1984, Mayflower moving vans — an Indianapolis-based company — hauled away the Baltimore Colts from the Charm City, where they had won championships with such legendary players as Johnny Unitas and Lenny Moore. They were bound for the Hoosier Dome, which had gone up before Indianapolis even had a team. Baltimore tried to keep the Colts via eminent domain, though it pursued its action too late, after the moving trucks had effectively removed the team from the city, so there was nothing to seize but memories. Maryland's solons enacted legislation permitting the city of Baltimore to seize the Colts. But Robert Irsay was too quick for them: the Colts were long gone by this time, and the city's would-be team-seizers were left grasping at air, rather like a prone and hapless defender watching Walter Payton whoosh by.

If you build it, they will come, right? Well, for once, the famous line from *Field of Dreams* came true. The Colts galloped out of Maryland under cover of night and into Indianapolis, whose city fathers and mothers were eager to show just how major league their city was. Sure, Indy had the 500 race, but that was only one day a year, and auto racing had yet to really break out of the gearhead ghetto. And the Pacers had been one of the four teams from the old American Basketball Association to be absorbed into the NBA in the merger of 1976–1977. But in the two *major* sports, baseball and football, Indy was not even an afterthought. Its AAA team in baseball, the Indianapolis Indians, was a farm club of the Cincinnati Reds, and even those major league owners who sought to line their pockets by pitting city against city tended to overlook the state of Indiana. But the NFL — now there was a shot worth taking. With only eight home games a year, as opposed to 81 in Major League Baseball, the stadium would surely sell out regularly.

The city whose detractors mocked it as "India-no-place" had built a $77.5 million downtown "Hoosier Dome" funded by a Marion County cigarette tax, a 1% food and beverage tax aimed at hotels and restaurants, and a gift of $25 million from the Lilly Endowment Foundation, an exemplary instance of local philanthropy.[44] The lease binding the Colts to the Hoosier Dome was 20 years.

The arrival — or theft, as Baltimoreans would say — of the Colts was expected to shower down prosperity like manna from heaven upon the capital of the Hoosier State. As one advocate breathlessly intoned, "If Indianapolis lands the Colts or any NFL team, it's going to do some amazing things for the city in terms of prestige, economic development, and in terms of enticing companies to locate to Indianapolis."[45] And there is no question that the Colts, especially after the arrival of star quarterback Peyton Manning, put Indianapolis on tongues that had never said the word before. But did the team really act as the blocking wedge for prosperity?

Indianapolis is usually held up as the primary — even sole — success story of subsidized sports. The city hatched a sports-fueled development strategy in the 1970s. The core of the strategy was to make Indianapolis a hub of amateur sports and to build the Hoosier Dome. In 1974 the city opened its $23 million Market Square Arena, which was home of the ABA/NBA's Indiana Pacers and a series of minor-league hockey teams until 1999, when the Pacers and, later, teams in the Women's NBA and Arena Football League moved into the $183-million Conseco Fieldhouse, also brought to the people of Indianapolis via taxpayer largesse.

During the period 1974–1992, as Mark S. Rosentraub, David Swindell, Michael Przybylski, and Daniel R. Mullins note in their survey of the city's revitalization efforts in the *Journal of Urban Affairs*, Indianapolis undertook eight sports-related construction projects totaling $172.6 million. They ranged from Market Square Arena to venues for tennis, bicycling, track and field, and the National Institute for Fitness and Sports. The centerpiece was the $77.5 million Hoosier Dome (later renamed the RCA Dome and finally demolished in 2008), which opened in 1984 to host the Indianapolis Colts. This strategy was pursued even though, as Rosentraub has pointed out, professional sports-related employment does not account for even as much as 1% of the private-sector payroll in any county in the USA.[46]

While sports projects constituted a minority of the downtown capital expenditures during this time, sports was "clearly an integral part of the overall development strategy for downtown Indianapolis."[47] So the quartet of authors compared job growth in the sports sector in Indianapolis with job growth in the sports sector in other cities; salary levels in Indianapolis with salary levels in nine cities with which Indianapolis is often grouped; and growth in Indianapolis with growth in other cities over the period 1977–1989. Not surprisingly, the boost in the number of sports-related jobs in Indianapolis was "quite impressive." Yet their impact on the city's overall economy was "inconsequential."[48] The sports strategy of Indianapolis "did not result in more growth than was experienced by other Midwestern communities and did not lead to a concentration of higher paying jobs in the region."[49]

After an exhaustive examination of the experience of Indianapolis, the city that sports subsidizers like to point to as their clearest example of success, Rosentraub concludes in another study, "Simply put, the sports strategy did not achieve its objectives." Sports-related jobs remained a tiny piece of the city's economy; the growth of the Lilly Corporation, "downtown Indianapolis's largest employer," was far more significant in stabilizing the city's job picture.[50]

The city and state do have a strong tradition of sports boostership, and the city's very conscious commitment to becoming a sports center would seem to make it the poster child for prudent subsidization. But Rosentraub says no: "Indianapolis's experience indicates that sports will not generate the growth or overall impacts its boosters and supporters frequently claim."[51] It just doesn't make that much of a difference — even in as concentrated a form as one finds in the capital of Indiana. In fact, "it is plausible to consider that, had the city focused on other factors, a larger economic impact would have been possible."[52]

And as Rosentraub, Swindell, Przybylski, and Mullins conclude in their study, "a sports strategy, even one as pronounced and as articulated as that of Indianapolis, is likely to have an inconsequential impact on development and economic growth" — no matter how many touchdown passes Peyton Manning threw.[53] The best example the pro-subsidy side has put up can't even make it — to mix sporting metaphors — to first base.

(For its part, the Hoosier/later RCA Dome, gone to that ring of hell reserved for indoor football stadiums, has been replaced by the $720 million Lucas Oil Stadium, which opened for gridiron business in 2008. The stadium was financed by a 1% tax on prepared food in nine of the surrounding counties, as well as an additional 1% restaurant food and beverage tax in Marion County on top of the tax already committed to paying off the late Hoosier Dome — which is gone but certainly not forgotten by taxpayers.)

The Teams by the Bay

In 1960, San Francisco unveiled Candlestick Park, new home of the relocated baseball Giants, who had spent their first two years in California playing in Seals Stadium, an old Pacific Coast League park. Candlestick was built for $15 million, with revenue from a bond issue as well as financing from the contractor, who was repaid from stadium revenues. The "Cave of the Winds," as it was popularly derogated, was an uncomfortable place to watch a ballgame — and sometimes even to play a ballgame. On July 11, 1961, at the All-Star Game in one-year-old Candlestick Park, fans across America were treated to the bizarre spectacle of relief pitcher Stu Miller of the National League being blown off the mound by a stupendously powerful gust.

No matter how much they huffed and puffed, the Giants ownership could not blow Candlestick down. And they did try. Three times between 1987 and 1990, San Francisco voters rejected public funding schemes to build the Giants a new ballpark. Meanwhile, regional voters in San Jose and Santa Clara also rejected three attempts to use their tax dollars to lure the Giants from their city by the bay.

San Franciscans, via a series of referenda, were actually able to pick out their preferred method of stadium financing. In 1987 they rejected Proposition W, which proposed to give the Giants choice land near the Oakland Bay Bridge and was mum on the question of how large a public subsidy the team would receive. Two years later, San Francisco voters narrowly rejected Proposition P, which called for a blend of public and private financing to build a new stadium in China Basin at an estimated cost of $115 million. They also rejected Proposition V, which called for publicly funded improvements to the woeful Candlestick Park. Meanwhile, in 1990, voters in San Jose and Santa Clara County rejected Measure H and Measure G, respectively, which would have put the taxpayers of those jurisdictions on the hook for publicly funded major league stadiums to which the Giants might move. Two years later, San Joseans again rejected a stadium measure, this time by a vote of 94,466–78,808, despite campaigning by such Giants greats, near-greats, and just-okays as Willie Mays, Willie McCovey, Orlando Cepeda, Vida Blue, Will Clark, Bob Brenly, and Tito Fuentes, and despite a lopsided spending disparity of $1.1 million for the pro-stadium side and $13,000 for the pro-taxpayer side.[54]

Finally, in 1993, the Giants hit the right note. By a margin of almost two to one, San Francisco voters approved Proposition B, which set aside the China Basin as the site for a new ballpark but called for no public funding and no taxes to be used toward its construction. San Franciscans wanted baseball — but they wanted those who owned and patronized the team to pay for it.

Giants owner Bob Lurie, frustrated that he could not stick his hand into the taxpayers' pockets, had announced in 1993 the sale of the team to a group of investors from Tampa. The league blocked that move. Why? Not least because, as deMause and Cagan write, such a move would have taken "Tampa Bay out of the running as a locale for other teams to move to — as several teams were then threatening to do."

So Lurie sold the team to a local syndicate, headed by Peter Magowan, which did what no owners ever do: they built the $357 million Pacific Bell Park (now AT&T Park) themselves. As deMause and Cagan write, the "San Francisco story…has become near-legendary among anti-stadium activists: the little electorate that could, calling the bluff of the leagues and getting to keep their team and their money, too."[55]

After a 1996 referendum expressing general support for a ballpark to be built via a private bond issue, seat licenses, naming and concession and advertising rights, and very limited public support was approved by almost two to one, ground was broken in the next year. The park, widely praised for such features as its picturesque use of the San Francisco Bay (into which mammoth home runs sometimes splash), opened in 2000. The owners obtained almost half ($170 million) of the park's funding via a private loan, another $70 million from charter seat licenses, $102 million from naming rights and sponsorships, and $15 million from San Francisco's redevelopment agency. That the most liberal city, politically, in America has a ballpark financed almost entirely by private enterprise is one of those twists that suggest that politics isn't all black and white.

Pacific Bell Park — or AT&T Park, or whatever it may be called two years from now — has been hailed as the "first privately funded ballpark built for Major League Baseball since Dodger Stadium opened in 1962."[56] It is also a testament to the stubbornness of the Bay Area voters.

The football 49ers, while meeting similar voter resistance to giveaways, have not been so fortunate. They would love to abandon Candlestick Park, even though the winds have never blown a quarterback over and the chill that drives fans to the concessions stand for coffee and hot chocolate seems fitting for football. The 49ers would gladly leave if they could — but they're having a tough time finding the door.

They thought they had a way out in 1997. Behind a war cry of "Build the Stadium — Create the Jobs!" a 1997 referendum calling for a $100 million city subsidy to a football stadium/urban mall project, fell apart in the wake of the multitudinous legal problems of then-owner Edward DeBartolo.[57] (Jim Ross of the Campaign to Stop the Giveway, the grassroots organization which fought against that 1997 referendum, said "It was beyond David versus Goliath. David at least had a sling. We were throwing rocks with our hands.")[58]

The 49ers resumed their push for a government-subsidized stadium in the twenty-first century. After fruitless haggling with the city of San Francisco, which, as with the Giants under Lurie, was much less willing to expend public funds on a stadium than are many putatively "conservative" cities of the Sunbelt, the 49ers asked the city of Santa Clara, 40 miles to the southwest, for a dance. Santa Clarans responded. In June 2010, Santa Clara voters approved Measure J, which called for leasing city-owned land to the 49ers for the construction of a new stadium. Although the Measure specified that no new taxes would be imposed on residents and that the city's general fund revenues would not be tapped for construction, the public share of the proposed $937 million stadium is $114 million. Or at least those are the numbers thrown around by the 49ers. Neil deMause, co-author of *Field of Schemes,* notes on this website (www.fieldofschemes.com) that in addition to the $114 million in direct subsidies, the Santa Clara plan calls for $330 million from "uncertain venue revenues" and a loan from the NFL that the NFL may not be in a position to make.

The 49ers spent $5 million promoting Measure J; anti-subsidy groups spent about $20,000. Yet the vote was much closer than the spending gap: Santa Clarans voted for the stadium by 11,231–7,609. A Santa Clara Stadium Authority is up and running, and if things go according to plan — and if the York family, owners of the 49ers, can't cut a more lucrative deal with San Francisco — ground will be broken on the stadium in 2012 with the goal of opening in the fall of 2014. Or maybe 2015. Don't expect to see the Santa Clara 49ers emblazoned across your sports page or the EPSN crawl, however: the name "San Francisco" says big league, while "Santa Clara" inspires puzzled head-scratching and a chorus of "Where is that?"

Throwing Money at Al Davis

In 1982, Al Davis, the owner and former coach of the Oakland Raiders who briefly served as commissioner of the American Football League in 1966, moved his Raiders to Los Angeles despite 13 straight seasons of sell-outs and a rich history by the Bay. John Madden, George Blanda, Jack Tatum: the Raiders were among the most colorful franchises of the old AFL and the post-merger NFL. But the city of Oakland and county of Alameda balked at installing luxury suites in the Oakland-Alameda County Coliseum, and Davis smelled revenues in the southland, so in March 1980 he declared that his team would relocate to Los Angeles to play in its rather more venerable Coliseum. The other owners promptly vetoed the move. Though NFL bylaws permit a team to move with the assent of three-quarters of the league's owners, those owners voted 22–0 (with five abstentions) against Davis. Consistent with his pirate style, Davis ignored the league and the Raiders made Los Angeles their home — temporarily, as it turned out. Oakland and Alameda County were left without a team but with annual $1.5 million debt payments on the Coliseum until 2004.

Oakland fought back in the courts. The Raiders filed an eminent domain suit, which failed. Interestingly, Rose Bird, the controversial California Supreme Court chief justice who became a symbol of Left Coast liberalism, cut to the nub of the matter: "The rights both of the owners of the Raiders and of its employees are threatened by the city's action...It strikes me as dangerous and heavy-handed for a government to take over a business, including all of its intangible assets, for the sole purpose of preventing its relocation."[59]

And in *Los Angeles Memorial Coliseum v. NFL* (1984), the Ninth Circuit Court of Appeals held that NFL rules circumscribing franchise movements were in violation of the Sherman Antitrust Act.

It should be noted that the league's principles are flexible. When the Los Angeles Rams announced in 1995 their intention of becoming the incongruously named St. Louis Rams, NFL owners rejected the move by a vote of 23–3, with six abstentions. But as Katherine C. Leone notes in the *Columbia Law Review*, the owners did an about face one month later, approving the Rams move by 23–6–1 after the team agreed to increase its relocation fee to the league from $25 to $46 million. It is merely a matter of sweetening the pot. As Leone notes, in the years since Al Davis moved his team without regard to NFL disapproval the league has shown "extreme reluctance to challenge relocations and a preference for capitulation over litigation."[60]

Davis found the grass no greener in Los Angeles, particularly in the dicey neighborhood of the Coliseum, which was also lacking in the luxury-suite department. So in 1987, Davis conducted a bizarre flirtation with the Los Angeles suburb of Irwindale. Irwindale officials and Davis agreed upon a $115 million football stadium into which the Raiders would move. Having deserted Oakland and its loyal fans, stepping out on the Coliseum was a snap. The Irwindale proposal fell apart, a victim of municipal politics and unrealistic expectations, but it boosted Davis's negotiating position as well as his reputation for knowing how to turn a buck. For the Raiders pocketed a nonrefundable cash payment of $10 million for even considering (and then discarding) Irwindale.

In 1995 the Raiders came home to Oakland — lured by $200 million in improvements by the Oakland-Alameda County Coliseum Authority. Yet, the leather-clad Davis continued to agitate for a new stadium, the implied threat being that he'll hit the road yet again if one is not forthcoming.

Over the years Davis has played any number of California cities for suckers. Not only Irwindale; he also toyed with Sacramento, which despite being the state capital has something of a second (or fourth) city complex. Sacramento pledged $122 million from city coffers to lure the Raiders when they were in Los Angeles. Gregg Lukenbill, managing partner of the Sacramento Kings of the NBA and a prime mover of the bid to bring the Raiders to town, said with King-sized hyperbole, "The Raiders moving to Sacramento would be an event of the magnitude of the Gold Rush."[61]

It didn't happen then, but never say never. The present home of the Raiders, O.co Coliseum, the new name of the former Oakland-Alameda Coliseum, opened in 1966, which makes it positively prehistoric by today's standards. The Raiders share it with the Oakland Athletics in the only baseball–football combination still using the same field. The Oakland-Alameda County Coliseum Authority Board, which oversees the stadium, is in the early stages of considering the erection of a new football-only stadium in the shadow — or maybe even in the footprint — of the current coliseum. Since the city and county are still jointly paying off the $20 million annual stadium improvements fee that helped convince Al Davis to bring the Raiders back to Oakland, local officials are adamant — so far — that taxpayer dollars will not build any new stadium for Mr. Davis's mercenaries. In 2010, Amy Trask, the Raiders chief executive, spoke excitedly of a "world-class stadium" on Oakland's horizon. It sounds like the taxpayers of Oakland are going to have a decision to make, sooner rather than later.[62]

A Modell City

The people of Cleveland could be forgiven for being somewhat jaded when it comes to owners threatening moves. The baseball Indians, who tired of cavernous Municipal Stadium almost before they moved into it in 1932, have made their share of threats. In 1984, county voters rejected a referendum on a 100% publicly funded downtown domed stadium for the Indians and the Browns of the NFL — despite the promise that 6,829 permanent jobs would result.[63] The teams stayed.

Six years later, Cuyahoga County voters narrowly approved (51.9%) a proposal to raise $169 million via 15 years' worth of alcohol and cigarette taxes in order to fund about half of the projected $344 million (a vast underestimate; cost was closer to $450 million) Gateway Center project, which included a new ballpark for the Indians (Jacobs Field; now Progressive Field) and a downtown arena for the Cavaliers of the NBA. Fans were badgered two days before the vote by this veiled threat from Major League Baseball Commissioner Fay Vincent: "Should the vote be a negative one, we may be finding ourselves confronting a subject that we want to avoid. I say to you, it would be very bad for baseball, and I am opposed to Cleveland losing its team."[64]

Sin taxes are an easier sell than, say, sales or property taxes, though, and the Indians got their ballpark. Tough luck for consumers of alcohol and cigarettes. But the Browns were left out in the cold — specifically, in the Lake Erie-chilled Municipal Stadium. And they wanted out — or at least up.

The Browns had played along the shores of Lake Erie for 50 years, featuring such greats as Jim Brown, coach Paul Brown — do you get the feeling the team's name fit superbly well? — Otto Graham, and Lou Groza. But Art Modell, who had purchased the team in 1961, was agitating for a taxpayer remodeling of Municipal Stadium. Cuyahoga citizens voted in November 1995 to tax themselves (or the sinners among them) $175 million to give Modell what he wanted. But the day before the election, the capricious owner announced that he was picking up his football (and the team that went with it) and moving to Baltimore, whose civic leaders had put together a sweetheart deal for a new stadium near the acclaimed new Camden Yards baseball park. (Cleveland's corporate and civic elite had conducted a $750,000 campaign to urge citizens to vote for the tax increase, which proposed to extend the levies on liquor and cigarettes and boost the sales tax by half a cent.)[65]

Modell had a 25-year lease that was supposed to keep the Browns in Cleveland through the 1998 NFL season, but the fine print in contracts is for the lawyers to haggle over. It can't prevent an owner from departing for greener pastures.

And those Baltimore pastures *were* greener: a "$200 million rent-free stadium with 70,000-seats, 108 luxury boxes, and 7,500 club seats," as well as a state-of-the-art training facility and *all* revenues from such diverse moneymakers as stadium naming rights, parking, and concessions.[66] The rent-free lease was for 30 years. Only a man truly loyal to a city and its fans could resist such an enticement, and Art Modell was not that man. By 1996, the Cleveland Browns had become the Baltimore Ravens.

"I am not about to rape the City as others in my league and others have done. You will never hear me say, 'If I don't get this I'm moving.' You can go to the press on that one. I couldn't live with myself if I did that."[67] So said Art Modell the year before he pulled the plug on Cleveland. Well, Art did live with himself, and as this went to press he had been living with himself for all these years — at a safe remove from Cleveland.

He left, as George Will wrote at the time, allured by "one of the most peculiar acts of government in memory" — Baltimore's throwing open of its treasury to Mr. Art Modell.

Typically absurd estimates of the economic impact the Baltimore Browns (soon to be renamed Ravens) would have on the area were disgorged by the usual suspects. The presence of an NFL team would pump $128 million into circulation in Baltimore, it was said, though as George Will points out, this was based in part on such "dubious assumptions...as that 20 percent of the people attending games will spend a night in a hotel" — this in a market whose verge already included teams in Philadelphia and nearby Washington. The state of Maryland's imaginative number-crunchers claimed that the team would create 1,400 full-time jobs, which, given the $250 million government subsidy, "comes to $178,000 per job," says Will, which is "rather pricey, even if the 1,400 number is not inflated and even if many of the jobs are not low-wage (ushers, food services) and seasonal."[68]

In all, the Maryland Department of Business and Economic Development claimed that a taxpayer-subsidized stadium for the Ravens would bring to the Charm City $111 million in economic benefits and 1,394 new jobs. (Don't you love that precision? Not 1,400 new jobs, as George Will had rounded off to, but 1,394. Such specificity gives the illusion of competence.) This department was running numbers for the governor; by contrast, the legislature's research service, the Maryland Department of Fiscal Services, estimated the bounty to be $33 million in benefits and 534 new jobs.[69]

Whether it would bring 1,394 new jobs, 534 new jobs, or no new jobs, to the city fathers of Baltimore, it was all worth it. The city was "rapidly moving back to being recognized only as the toilet stop on the drive between Washington, DC and Philadelphia," as one wag put it.[70] Rejoining the NFL would be like dressing Baltimore in the regalia of the big-time city once again.

Art Modell proved that "there often is no penalty for failure in America," observed George Will, especially if he who fails has connections in high government places.[71]

"No team, no peace," cried Cleveland Mayor Michael R. White. The mayor had supported the tax hike to propitiate Modell because, he claimed, the Browns pumped $47 million annually into the local economy.[72] A flood of lawsuits were filed against Modell and the Browns. The league, realizing it had a public relations disaster on its hands (the Browns, though consistently a poor team in recent years, were a storied franchise who drew over 70,000 fans per game even in lousy years) and not too certain that the legal obstacles could be surmounted quickly enough to guarantee Baltimore would be hosting a team in the coming season, struck a deal with the city. Modell's gypsies moved to Baltimore and became the Ravens, a poetic nod to Edgar Allan Poe. But unlike such uprooted teams as the Rams and Colts, the Browns' history stayed in Cleveland. The team's colors, uniforms, name, logo, and records stayed in their native city.

Cleveland razed Municipal Stadium, to the accompaniment of few tears. The public share of the $283 million Cleveland Browns Stadium, which opened for the revived Browns franchise in 1999, was 70%. Art Modell had fled town, but the Browns got their stadium anyway.

The Florida Bridesmaid

The legendary Chicago newspaper columnist Mike Royko wrote in 1982, "If there is anything that marks a town as being a genuine hicksville, it is the innocent belief that a domed stadium is the height of progress."[73] By that measure, nowhere was as hicksville in the 1980s and 1990 as the neighboring and rivalrous Florida cities of Tampa and St. Petersburg.

Separately and together, Tampa-St. Petersburg was the perennial bridesmaid of baseball — though it was a bridesmaid who helped many a groom strike a rich bargain with the bride. As Bob Andelman writes in *Stadium for Rent: Tampa Bay's Quest for Major League Baseball* (1993), "no community risked as much or embarrassed itself as greatly as did the City of St. Petersburg in its pursuit of a Major League Baseball team."[74] For about a five-year stretch from the late 1980s through the early 1990s, Tampa and St. Pete (separately at first, together later) tried to lure the Chicago White Sox, Minnesota Twins, Texas Rangers, Oakland Athletics, Seattle Mariners, San Francisco Giants, and the teams that became the Florida Marlins and Colorado Rockies. (For a detailed account of the sometimes desperate attempts by Tampa and St. Pete to attract a Major League Baseball team, see Andelman's *Stadium for Rent.*)

When the White Sox made noises about departing Chicago, St. Petersburg jumped into the pool of suitors with an extraordinary splash. The city council of this city of under 250,000 — that is, less than one-tenth of the population of Chicago — voted to build an $83 million domed stadium even though the city had no assurances — not even a wink–wink, hint–hint — that the White Sox were coming.

Tampa, 20 miles away across Tampa Bay, had entered the big leagues in 1974, when the city won the dubious prize of a Buccaneers' football team that began playing in the NFL in 1976, only to lose all 14 of its games that first season. Tampa coveted Major League Baseball, too, but its sister city, St. Petersburg, coveted it more. The cities squabbled, quarreled, backstabbed, and undercut each other in pursuit of a Major League Baseball team until St. Pete's gamble — the Suncoast Dome — relegated Tampa to sputtering and then support for what was now envisioned as a regional team.

The dome was under discussion for much of the 1980s, and the one group that was barred from an effective role in the discussion was — naturally — the taxpayers. Citizens demanded a countywide referendum. Many were aghast at the thought of spending up to $100 million on a dome that had, as yet, no tenant. St. Pete officials insisted that if you build it, they will come, but the problem was that many of the "you"s did not want to be bled of tax money in order to build it. As Bob Andelman wrote, "elected officials knew a referendum would never pass. That's why it was never offered." Pinellas County commissioner Bruce Tyndall, a critic of the dome, said, "It boiled down to the use of public dollars for what I would call private enterprise.... [T]he supporters weren't putting any money on the line. The question was, would I have voted my family dollars in support of it? It's a lot easier spending somebody else's money than your own. That was the philosophy I used."[75]

As Andelman notes, by 1984 St. Petersburg had already demolished 264 buildings and relocated 461 households and bought 66 acres of land, all at a cost of $13 million, before a single spade had been turned on this project. The estimated $85 million was to be raised by a hotel tax (40% of the total) and sales tax (60%), which are often regarded as politically painless ways of raising funds. (Jacking up property taxes gets politicians defeated and is absolutely verboten — unless there is no other choice.)

St. Petersburg won admiration for its sheer doggedness, however unwise it may have been. Peter Ueberroth, commissioner of Major League Baseball, even sent the city leaders a telegram in 1986 with a remarkable warning:

> You have before the citizens of St. Petersburg important decisions regarding construction of a new stadium. On behalf of Major League Baseball, I want to reaffirm to you that no assurances can be given with respect to the establishment of a Major League Baseball franchise in St. Petersburg. Indeed, in our evaluation of potential cities for relocation or expansion, St. Petersburg is not among the top candidates. We do not want there to be any misunderstanding of baseball's position. You must recognize that any current decision by St. Petersburg to undertake the construction of a facility capable of housing Major League Baseball will be made by your community without any encouragement whatsoever on the part of Major League Baseball.[76]

The commissioner could not have made it any plainer. If you build it, we probably will not come. But the city went ahead. Bob Andelman quotes retired city judge Henry Esteva as telling a city council meeting, "The city is the bride, the county is the groom, the stadium the unborn child. There's a beautiful courtship. The city and county get together despite opposition from family. They get engaged, then they get married in a beautiful ceremony. Even before the honeymoon is over, there is trouble. A jealous monster across the bay [Tampa] does not want the couple to have a child. But they hang on. The wife becomes pregnant and the child is nearing delivery. Even some relatives of the wife want her to abort. Let me tell you something. The delivery room is too late for abortion!"[77]

This was a bizarre analogy. The unborn child, for one thing, was the product of an artificial insemination by the taxpayers of Pinellas County — not exactly an immaculate conception, or even a love child! But council went ahead with the Suncoast Dome, which in time became Tropicana Field, and the monster across the bay was defanged. St. Petersburg councilman Bob Stewart said, "We just hope history will prove we were visionaries and not horses' derrieres."[78]

It took several more years and several more heartaches before the empty dome, which wound up costing $130 million, had its team. In 1991, St. Petersburg was confident that it would be named one of the National League's two new expansion teams, but those franchises went instead to Miami (the Florida Marlins) and Denver (the Colorado Rockies). A disappointed booster complained, "We've been used as a nuclear threat to other communities to make them give teams whatever they want."[79] But still, St. Pete kept at it. In 1993, it gained a temporary tenant in the Tampa Bay Lightning of the National Hockey League, which played in the dome for three years. In 1995, Major League Baseball finally awarded an expansion franchise to the region — the team to be called the Tampa Bay Devil Rays, to the chagrin of St. Petersburg patriots. Tropicana bought the naming rights to the dome, which underwent a $70 million fix-up job before the Devil Rays took the artificial field in 1998.

The taxpayers built it, and the majors did come. Despite the inclusion of a rotunda in homage to Ebbets Field, ballparkdigest.com rated Tropicana Field the second-worst park in the majors — second only to the Metrodome in which the Minnesota Twins played.[80] Now that the Metrodome has gone to that junkpile in the sky, Tropicana Field, the stadium that dreams (and taxpayers) built, is the worst park in the majors. Congratulations, St. Pete!

Chicago Shakedown

Mike Royko ("If there is anything that marks a town as being a genuine hicksville, it is the innocent belief that a domed stadium is the height of progress") pegged Tampa-St. Petersburg with pinpoint accuracy. The Chicago White Sox used the Bay Area of Florida — which was acutely sensitive to the fact that the Bay Area across the country, that of San Francisco-Oakland, had four teams in Major League Baseball and the National Football League — with the manipulative skills of a master sadist.

White Sox owners Jerry Reinsdorf, an attorney, and Eddie Einhorn, an executive with CBS, had bought the Pale Hose in 1981 from the legendary Bill Veeck. The White Sox still played in Comiskey Park, which the notoriously tight-fisted but beloved Charlie Comiskey had built in the style — appropriately modified — of the Colosseum in Rome with his own money back in 1910.

Owner-eponym Charlie Comiskey built the park for $700,000 — about $150,000 for land, $550,000 for construction, and *all of it* financed privately. Comiskey commissioned architect Zachary Taylor Davis to design the park — a "baseball palace of the world," as it became known. (A couple of years later, Davis would draw up what became Wrigley Field.) Comiskey, whom even his friends admitted was a cheapskate, was heartily disliked by White Sox players — for a taste of their distaste, see John Sayles's movie *Eight Men Out* (1988), which is about the 1919 Black Sox scandal in which several White Sox players threw the World Series. But the fans loved him. As Michael Benson writes, "Any worthy Chicago organization that wanted to use his park for an event had it for free — assuming they were White Sox fans, of course, and they always were." Comiskey is said to have remarked, "The fans built the park, didn't they?"[81] While they didn't — *he did* — later White Sox owners would have cause to make the same observation.

By the early 1980s, Comiskey's neighborhood was largely African-American, and though the team drew well when it played well — as in 1983, when the White Sox made the playoffs — attendance slumped during the down years. The White Sox were not like the Cubs, lovable losers. When Chicago's second team lost, the fans stayed away.

White Sox owners Jerry Reinsdorf and Eddie Einhorn were fishing about for a new stadium. Did they do as Charlie Comiskey had done, and put together a land acquisition-stadium construction package in the private market? Of course not. Instead, they buttonholed their old law school classmate James Thompson, who by a most fortunate coincidence happened to be the Republican governor of Illinois. Reinsdorf says that the governor told them on Opening Day 1985, "You'll never get one built unless there's a crisis. Unless people think you're going to leave if you don't get one."[82]

Wise, if cynical, advice.

And Governor Thompson was playing it both ways. "I'll bleed and die before I let the Sox leave Chicago," he said in 1988, overdosing on the self-dramatics.[83]

The owners eagerly took Governor Thompson's advice. Explained Jerry Reinsdorf: "We had to make threats to get the new deal. If we didn't have the threat of moving, we wouldn't have gotten the deal."[84] They dallied with suburban Addison, Illinois, and Tampa, St. Petersburg, Phoenix, Denver, Washington, and Jacksonville.

Addison voters rejected a proposed White Sox stadium in a nonbinding referendum in 1986, thanks to the combined effort of "homeowners and environmentalists." The Sox outspent the anti-Sox side $100,000–$3,000 in Addison, but lost by a hair's breadth.[85] Reinsdorf blamed "anti-zealots," which is another way of saying "taxpayers who don't want to give me money."[86] (The taxpayers are happy to return the fire. The acronym of one later anti-Chicago Bears stadium subsidy group was STINCS, or this Stadium Tax is Nothing but Corporate Subsidy. Other groups in other cities are free to borrow the name, whose application seems more or less universal.)

Owners Reinsdorf and Einhorn, after a rebuff from their favored suburban location in Addison, Illinois, took another look at Chicago. Mayor Harold Washington put together a proposal under which the White Sox and Bears would move into a new stadium on Chicago's West Side "to be financed with $255 million of industrial revenue bonds and subsidized by county and state support."[87] This stadium was never built, in part due to community opposition from such groups as the Interfaith Organizing Project, which called the park "a playground for the wealthy." In fact, perhaps the word "opposition" is too mild for what some residents felt. After Mayor Washington died of a heart attack in 1987, and the West Side stadium died, too, the Reverend Arthur Griffin of the Interfaith Organizing Project attributed the mayor's death to his support of the stadium: "It was a base and immoral decision that represents the antithesis of all that is decent and humane, and so [Mayor Washington] was soon removed. If acting Mayor Eugene Sawyer persists…he too will soon be removed because 'God don't like ugly' and this pitiful display of hedonistic pleasure for the rich at the expense of the poor is ugly, ugly, ugly."[88]

That is one wrathful minister talking — or one wrathful taxpayer. In any event, Harold Washington never built his stadium for the White Sox. Instead, the Sox owners opted to knock down Comiskey and build a new stadium across the street — in the process stealing via eminent domain 178 private homes and a dozen businesses.[89] Chicagoans fought to save Comiskey under the banner of "Save Our Sox," but it was fruitless. The baseball palace of the world was reduced to rubble. Eighty years of White Sox history, of Shoeless Joe Jackson and Bill Veeck and the Go–Go Sox of 1959, were demolished to make a parking lot which would serve a government-subsidized structure lacking any history save its sordid conception in backroom Illinois political deals.

The Illinois legislature, never exactly a model of probity, rectitude, and high honor, passed legislation authorizing the construction of a new park for the White Sox, to be paid by state bonds and a hotel tax. The deal was worth $150 million, and the White Sox got a subpar park just a Ron Kittle cannonshot away from old Comiskey. The lease for New Comiskey — since renamed U.S. Cellular Field — is with the state of Illinois, and the White Sox pay rent in the grand total of $1 annually.

Charlie Comiskey liked a shrewd deal as much as any man, but it's probably for the best that the team took his name off the new ballpark. Charlie Comiskey paid his own way.

Joe Robbie Shows the Way

The shining exception to the subsidized murk is in Miami, of all places, where the $115 million Sun Life Stadium, nee Joe Robbie Stadium in 1987, home of the NFL Dolphins, the MLB Marlins (until 2012), and NCAA Miami Hurricanes, is the showcase private multipurpose stadium.

It is not that Dolphin management was so pure of heart as to nobly refuse preferment from state and local governments. Rather, it is that the voters were so wise as to refuse to give the footballers a handout. Over the span of 12 years, Miamians thrice rejected bond referenda to finance a playing field for the Dolphins. So Joe Robbie, the owner, a self-made man from South Dakota, decided to build the thing himself — a "watershed in stadium economics."[90]

The Dolphins had played in the Orange Bowl dating back to their days swimming in the old AFL. Joe Robbie wanted a state-of-the-art stadium, however, and something less cramped than the Orange Bowl, which squeezed its patrons in. Robbie financed the stadium in part through his revolutionary use of luxury suites and personal seat licenses, which ordinary fans may bewail but which have the virtue of transferring the burden of paying for stadium construction from the taxpayer to those who actually use the venue. After all, one study found that the median income of purchasers of tickets to pro sports events was 84% above the median income level, which means that stadium subsidies amount to a redistribution of wealth upwards.[91] Not only in the obvious sense of rich owners and players sharing the booty, but also in form of a subsidy of affluent ticket buyers by less affluent average taxpayers.

Joe Robbie Stadium opened in 1987. With its 183 luxury suites and 10,209 club seats, it showed other NFL owners how a stadium might be privately financed and profitable. They didn't listen. Oh, they made extensive use of personal seat licenses and luxury suites, but these were to be *in addition to*, not in place of, government subvention.

For another $10 million, Joe Robbie was easily adapted to use by the baseball Florida Marlins, who were not born until 1993.

Unfortunately, the late Joe Robbie's example has not proven instructive to the current owners of the Florida Marlins. The baseball team is leaving the privately constructed confines of Sun Life Stadium for public accommodations.

For the 2012 season the Marlins moved into the 37,000 seat, retractable-roof-covered Marlins Park, which was put over on the people of Miami-Dade County by what Jeff Passan of Yahoo! Sports calls "baseball's biggest welfare case" — the Florida (soon to be Miami) Marlins, owned by the stupendously wealthy art dealer Jeffrey Loria.

Loria and the team's president, David Samson, had pleaded if not poverty then barely keeping-our-heads-above-water status for several years in an attempt to get local governments to build them a new playpen. The plan worked, when in 2008 several years of negotiations (including veiled threats to move) resulted in Miami-Dade County Commissioners approving a $634 million baseball-only ballpark whose primary funding ($409 million) is in the form of loans "loaded with balloon payments and long grace periods" which will, when the last is finally paid off in 2049, have cost $2.4 billion.

A circuit court judge ruled that voters had no right to force a referendum on the deal, so the deal slid through. Marlins Park rose, courtesy of Miami taxpayers. Yet documents obtained by the website Deadspin reveal, in Passan's words, that "the Marlins could have paid for a significant amount of the new construction themselves and still turned an annual operating profit. Instead, they cried poor to con feckless politicians that sold out their constituents." Through accounting legerdemain, the team essentially hid $48.9 million in profits in 2008–2009 so as not to appear capable of contributing any more than the minor share ($155 million) of Marlins Park for which they are responsible.

Passan notes that in "the annals of bad stadium deals, it's among the most odious, right alongside the Washington Nationals' extraction of $611 million from the DC city council to get Nationals Park built." But that's Washington, after all; we expect waste to flower like cherry blossoms along the banks of the Potomac. Miami had Joe Robbie's example to guide by — and it steered in exactly the opposite direction.

Adding insult to injury, the Marlins even finagled $5.3 million from the Miami-Dade Art in Public Places program. The Marlins — that is, the taxpayers — forked over $2.5 million to artist Red Grooms to "design a piece with pelicans and seagulls and bright colors and abstract shapes and, best of all, animatronic marlins that celebrate home runs."[92] As Phil Rizutto would yowl, "Holy Cow!"

At least the team will finally shed its presumptive statewide tag and be known as the Miami Marlins.

Meet Them in St. Louis

The unimaginatively named St. Louis Cardinals of the NFL began life in 1920 as the Racine Cardinals, though by 1922 they had moved to Chicago, where for almost four decades they played second banana to the Bears in the Second City. In 1960, owner Violet Bidwell moved the team to St. Louis, where it shared a stadium with the much-better-known Cardinals of baseball's National League. The franchise had been stuck in the doldrums since Otto Graham was a tyke, or so it seemed.

The Bidwell family (Violet's sons Charles and William had inherited the team upon the matriarch's death) started sniffing around for new cities and better offers as early as 1964, when the city of Atlanta made a pitch, but St. Louis, with some assistance from Anhauser-Busch, built multipurpose Busch Stadium, into which both Cardinals teams moved in 1966. The Bidwells stayed put. For a bit.

Busch Stadium was a key piece of the downtown revitalization strategy in the St. Louis of the 1960s. Like other government-subsidized stadiums of the 1960s, it was sold to voters as a wellspring of prosperity. Yet as George Lipsitz of the University of Houston at Clear Lake City wrote of St. Louis and kindred cities in the *Journal of Sport and Social Issues,* "Voters in each city approved revenue bonds for land clearance and construction in anticipation of widely distributed benefits that never materialized."[93]

St. Louis is a case study in the way that powerful interests successfully cloak their raids on the treasury in high-minded rhetoric and unimpeachably liberal platitudes. George Lipsitz has described the process by which Busch Stadium was conceived and delivered by Civic Progress, an organization of St. Louis corporate leaders formed in 1953 to serve, in the sanctimonious words of Mayor Joseph Darst, as "the conscience of the community." The group's rather strange motto — "We expedite" — promised fast action, and in fact it served as a community catalyst for such dubious blessings as highway construction, urban renewal, and slum clearance. As George Lipsitz points out, the demolition of "slums" in St. Louis "left almost 30,000 poor people homeless and the city never developed adequate plans to relocate them."[94]

Among the corporate chieftains of Civic Progress was August Busch Jr., brewing heir who purchased the baseball Cardinals in that same year of 1953. The entry on Busch in the *Dictionary of Missouri Biography* quotes a sportswriter providing this assessment of Busch's reputation before he bought the Cardinals: He "chased women right and left...[was a] heavy, two-fisted drinker and was disliked by more polite society...People just put up with him. He had no reputation at all as a civic leader."[95] But upon purchasing the Cardinals with his inherited money (his grandfather had founded the brewery), Busch employed a public relations firm to burnish his image.

By the 1960s, Busch wanted a new ballpark for his team. The Cardinals played in Sportsman's Park, a late-nineteenth century venue that had been extensively renovated in the early twentieth century and had served as home to the Cardinals and the baseball Browns, before they flew the coop to become the Baltimore Orioles. Busch had renamed Sportsman's Park "Budweiser Park" upon buying the team, but this was too much for other National League owners, who demanded a name change. So Busch put his own surname on the park. Sportsman's Park had a "garish, county-fair sort of layout," according to legendary sportswriter Red Smith, but this would not do in modern times: a bland cookie-cutter stadium would be so much better![96]

Placing a new stadium in downtown St. Louis became a high priority for Civic Progress. Federal funds were showering all over the city in the 1960s. Highways were being built (and neighborhoods destroyed), a convention center and mall rose up: city leaders were looking to Washington, DC, for rescue, and the lifeline (or noose?) was extended. The feds even laid the groundwork for the new Busch Stadium. Writes Lipsitz: "Civic Progress succeeded in getting the area for the proposed stadium declared 'blighted' and consequently availed itself of opportunities to form a redevelopment corporation that could acquire land for a stadium through powers of condemnation and eminent domain established for urban renewal and slum clearance purposes."[97] Yet another field of dreams was to be built on land seized from the less fortunate.

Small business owners who were to be kicked out of their shops and buildings in order to accommodate the Cardinals protested, to no avail. They objected that it wasn't fair or just to relocate the little guys to make way for the big guys, but the little guys failed to convince their neighbors. In 1962, St. Louisians approved $6 million in public improvements to the area that would host the new park, to be funded through a bond issue.

The baseball Cardinals kicked in $5 million toward construction of the $20 million Busch Stadium, which opened in 1966 and hosted both the football Cardinals and the baseball Cardinals. The last baseball game was played on its artificial surface in October 2005, and a month later the park was demolished. The Cardinals moved into a new venue called, naturally, Busch Stadium, whose $365 million cost was financed largely by the team, though it did include a $45 million loan from St. Louis County and a 25-year exemption from property taxes on the stadium.

As for the football Cardinals, the Bidwells claimed that the 55,000 seat Busch Stadium was too small, even though the team rarely sold out. The Bidwells kept up their whining — but not their winning — as well as their threats and in 1987 decamped to Phoenix when St. Louis refused to build the team a $120 million stadium. The relocated Cardinals played in a college stadium (home of the Arizona State Sun Devils) till the state-of-the-art modernist facility now known as the University of Phoenix Stadium opened in 2006. Of its $455 million cost, the Cardinals contributed less than one-third ($143 million), with the bulk being the responsibility of taxpayers in the form of the Arizona Sports and Tourism Authority ($302 million).

(Making a parenthetical digression to the southwest, Phoenix is on a roll, corporate-welfare-wise. Its Arizona Diamondbacks baseball team, born in 1998, plays at Chase Field, formerly Bank One Ballpark, a $349 million project to which the people of Maricopa County, perhaps against their will, contributed $238 million via a quarter-cent sales tax. Citizen-taxpayers, outraged at this bald subsidy, organized, seeking a referendum. Polls showed that the stadium tax would lose if put to a vote. One creative activist flooded the county's board of supervisors with postcards demanding that the stadium be named the "Maricopa Tax Dome." Citizens were furious — and they were powerless. The state legislature had granted Maricopa County taxing power, "as a legal subdivision of the state," to raise the money. So the tax increase was not subject to referendum. The usual pie in the sky economic projections were made that the ballpark would rejuvenate downtown Phoenix, which was suffering through a slough due to an abundance of vacant downtown office space. That hasn't happened yet, but the people of Maricopa County do have "one of the most elaborate yet ostentatious ballparks ever conceived," as Richard Temple Middleton IV writes.)[98]

Now back to St. Louis.

The eccentric Georgia Frontiere, who had inherited the Los Angeles Rams from Carroll Rosenbloom, a man with dubious acquaintances who had drowned in 1979 under *very* mysterious circumstances, had run that franchise into the ground and was looking for a way out of Anaheim, to which the Rams had moved from the Los Angeles Coliseum in 1980. (Carroll was Georgia's sixth husband.) St. Louis beckoned, flashing a wad of cash. To lure the once-proud Rams franchise, St. Louis built, entirely on the public nickel, the Trans World Dome, now known somewhat prosaically as the Edward Jones Dome, a sterile 66,000 seat facility constructed for $280 million by the St. Louis Regional Sports Authority. There was not even a cursory effort to pretend that this was a private–public partnership. The leadership of St. Louis was of the opinion that losing the football Cardinals had badly dented the city's image, making it appear borderline major–minor league. It had no NBA team, after all, so it was present in only two of the country's four major sports. The Rams were presented not only with a key to the city but also a key to the treasury, as St. Louis even covered $15 million in relocation costs and retired the Rams debt ($30 million) to Anaheim, the city the franchise was abandoning for the arches of St. Louis. The Rams, who subjected St. Louis fans to four bad years of football before quarterback Kurt Warner led them to a sudden and unexpected prominence, need not have worried about filling the new dome, either: St. Louis had guaranteed a per game ticket sale of at least 55,000, with the city buying tickets itself, if necessary.

The Trans World Dome opened in 1995 to house the relocated Los Angeles Rams, and within a decade and a half it was derided as obsolete. The public pumped another $30 million into a 2009 renovation, gussying up the turf and the scoreboards and even the paint inside the dome. But the team's lease with the St. Louis Convention and Visitors Commission, operators of the Ed, as it is familiarly called, permits the Rams to look elsewhere for fields on which to play if the dome is not in the top 25% of all NFL stadiums, as measured by 15 conditions ranging from seating to luxury suites to locker room facilities to lighting.

Never mind keeping pace with the Joneses; in St. Louis, the Jones Dome has to keep pace with the rest of the NFL.

Seeing the Future in LA

With the Rams off to St. Louis and the Raiders back in Oakland, the second-largest city in the USA has been without a National Football League team since 1995.

Certainly Rams fans remain bereft, as do those (outside of the California prison system) who took the Raiders to heart in their Southern California sojourn, sandwiched between Oakland residences. But many Los Angeles fans are happy to have a choice of games on Sunday rather than be force-fed the afternoon contests of what would be, at first, a lousy expansion team.

The lack of any burning desire on the part of Los Angeles football fans for a new NFL team hasn't kept the owners of perhaps a dozen or more other teams from threatening, either explicitly or via winks and nods and meaningful looks, to pack the moving van and decamp to the land of perpetual sunshine.

Majestic Reality, a Los Angeles real-estate firm with close ties to LA Mayor Antonio Villaraigosa, has proposed constructing an $800 million stadium in the grimily named City of Industry. The city's name is a splendid example of truth in advertising: Industry is all about industry. It has 777 residents but over 2,500 businesses providing over 80,000 jobs. With no business tax but plenty of strip clubs, Industry is a let-it-all-hang-out kind of place. And it wants, improbably, a pro football team. In 2009, its tiny electorate approved the issuance of $500 million in bonds to pay for infrastructure to support a proposed Los Angeles Football Stadium, which would be a privately financed project of Majestic Realty. *Field of Schemes* author Neil deMause told leftist sports journalist Dave Zirin about the plan: "It's a weird one, in large part because the City of Industry is so weird. Arnold's [then-Gov. Schwarzenegger's] claim that it's entirely privately financed is a crock — the land and infrastructure is being funded by property taxes — but in a town with barely any actual people in it, you could legitimately argue that the industry in Industry is just voting to tax itself to bring the NFL to town."[99]

The Industry project has yet to get a green light, though not for lack of effort. According to *Capitol Weekly*, the magazine of California politics and government, Majestic chieftain Edward Roski contributed $700,000 between 2007 and 2009 to various candidates and causes, most of them allied with Mayor Villaraigosa.[100]

Though Majestic does not have its stadium — and the 40% ownership stake of the Los Angeles team it is demanding as a condition of building Los Angeles Football Stadium — it has done the NFL a considerable favor. In January 2010, Majestic managing partner John Semcken named seven franchises — the Jacksonville Jaguars, Buffalo Bills, San Francisco 49ers, San Diego Chargers, Minnesota Vikings, St. Louis Rams, and Oakland Raiders — as possible relocators to the sunny climes of Industry. He singled out Jacksonville and Buffalo — "two teams in very, very small markets...that have either outdated stadiums or are having trouble filling their stadiums or both" — as his prime candidates.[101] (The fact that Buffalo, whose Ralph Wilson Stadium was paid for almost entirely by taxpayer dollars and whose fans annually pack the stadium to watch a frequently bad team, is on the list shows that no act of loyalty or suckerdom ever goes unpunished.)

Even if none of these teams ever picks up stakes and takes off for the brown grass of Industry, the threat will have served a purpose. Each of the teams named will be in a better position to wheedle and cajole and coax additional subsidies from their current host cities.

Los Angeles may not have a team, but it remains, in a very real sense, in the league.

Houston, We Have a Problem

Texas came late to the major league party. But when it did come, people noticed.

Houston entered baseball's National League in 1962 as the Colt .45s — a tribute to the gun, not the malt liquor. The team played three seasons at the hastily built and temporary Colt Stadium before moving into what the Reverend Billy Graham called "in truth one of the great wonders of the world": the Astrodome.[102]

Judge Roy Hofheinz was the man who brought the Astrodome into being. The introduction of a single fact — that the Judge was a close associate of Lyndon B. Johnson — should tip off even the most obtuse observer to the likelihood that the Astrodome's lineage is not one of pure free enterprise — even though, as George Lipsitz sardonically noted, the school board of Houston was at this same time refusing participation in the federal school lunch program because it smacked of "socialism."[103]

According to a 1968 profile in the *New York Times Magazine,* in 1936 Roy Hofheinz became, at age 24, the youngest county judge in America (Harris County, Texas) due to the assistance of a young politico named Lyndon Baines Johnson. The Judge returned the favor by managing LBJ's first Senate campaign. Like LBJ, the Judge made a fortune in radio. (It helps to have an in with the dispensers of licenses.) He also served two terms as Houston's mayor and, in the manner of Texans of his and other eras, he dreamed BIG. Houston, he believed, was a major league city. And a major league city deserves a Major League Baseball team. (The belief that having a team in the highest leagues certifies a city as major league is remarkably durable. Twenty years later, an Oakland attorney, contemplating the departure of the NFL Raiders to Los Angeles, asked, "Have you heard of Brooklyn since the Dodgers left?")[104]

When the National League seemed receptive to Houston's bid for a franchise, Hofheinz and his allies in Houston's oligarchy went to work. His associates included, prominently, oilman-rancher Bob Smith, whose fortune was estimated at between $100 and $500 million by *Fortune* magazine in the mid-1960s, and Craig Cullinan Jr., grandson of the founder of Texaco. So what did this trio of financial power-hitters do? Did they build an Astrodome as a monument to the oil upon which so many Texas fortunes have been made? No. Instead, the "wealthy citizens of Houston promoted a county bond issue authorizing construction of a major-league stadium."[105]

Houston was major league, all right — and the citizens of Harris County were being called upon to prove it.

Smith and Cullinan led the bond campaign, which succeeded. The Astrodome was built with $35 million of taxpayers' money raised through bonds and also taken from the general fund. (A total of $31.6 million came from the county, and the other $3.7 million was funneled through the city of Houston.) The Houston Sports Association, whose leading stockholder was Texaco heir Cullinan, contributed $6 million "to construct and decorate the skyboxes" in which, coincidentally, its high-rolling partners would be sitting.[106] Hofheinz made sure that he had a "palatial suite and a private bowling alley built for himself" in the Astrodome — apparently he believed that taxpayers should subsidize private keglers as well as major league sports.[107]

Roy Hofheinz, known simply as the Judge, did much of the political spadework for the Astrodome in a manner that would have made LBJ proud. According to the *New York Times*, "he leaned on the Texas Highway Department to move ahead by five years its plans to build a 14-lane expressway bordering the stadium site." He also conjured up $750,000 "from the Federal Government by promising to equip the stadium as an emergency fallout shelter." Even the Cold War could be put to use by an opportunistic seeker of federal stadium funds. (Not that the Judge was an impeccable analyst of market trends. While he controlled a North American Soccer League team and a minor league hockey team that played to relatively sparse crowds at the Astrodome, he turned down a chance to buy into the American Basketball Association because, as it was explained in 1968, "he doesn't believe pro basketball will sell in Houston." Ask the Rockets about that one.)[108]

The Astrodome was a marvel of modern engineering and a model of sterility. The temperature was a constant 72°. The wind blew at one-mile-an-hour, not nearly enough to help Houston players hit homers in what was a notoriously unfriendly park to hitters, with its 390-foot power alleys. No dome game is ever rained out, and domes, unlike open-air parks, can easily host non-sporting events: RV shows, comic-book conventions, and the like.

Yet domes are also boring. In shielding spectators from the elements they remove one of the elements that makes spectating enjoyable: the sense that one is sitting in the great outdoors, breathing fresh (or relatively fresh — the Meadowlands excepted) air, and watching superbly skilled athletes. So folks in Houston grew blasé about the Astrodome. And even a $67 million expansion in 1989 was not enough to save the eighth wonder of the world from obsolescence.

While the Astrodome was not blasted to that domed stadium graveyard in the sky — instead, it sits alone and mostly empty and rather pathetic — it was abandoned, first by the Oilers, who split for Tennessee, and then by the Astros. Replacing it for the 2000 season was the unfortunately named Enron Field, naming rights purchased by the scandal-ridden Houston-based energy company. As soon as could decently be done, the field was rechristened Minute Maid Park. Its $250 million construction cost was borne largely by the taxpayers, though the people of Harris County, Texas, tried to spread the pain by slapping a 2% hotel tax and 5% rental car tax on those who take lodging or rent a vehicle in the county. The taxes, which raised $180 million, or almost 70% of the cost of the stadium, were approved in a 1996 referendum which was conducted under a very large shadow: that of the football Oilers relocating to Tennessee, where they became the Titans. Houstonians did not want to lose their franchises in the two major sports leagues, and so with a promise from Astros' owner Drayton McLane Jr. that if the voters approved the taxes the team would stay, the voters said yes. Of course they were operating under the illusion that they were taxing outsiders: the people who stay in the 400 or so hotels, totaling 55,000 rooms, in the county, and those who rent cars within the district. But the Astros stayed.[109]

Unlike the Astros, the football Oilers skipped town.

When owner Bud Adams was threatening to move the Oilers out of the Astrodome and into a city more amenable to his charms (or responsive to his begging), Houston Mayor Bob Lanier resisted Adams's request for $150 million, give or take a few mill, asking, "Can you ask the average guy to build luxury suites for rich people, so they can support rich owners, so they can pay rich players?"[110] Mayor Lanier also lamented that "teams have made hostages of urban taxpayers," and it was doubly unfortunate that urban politicians were willing — even eager — to negotiate with the hostage-takers.[111]

Mayor Lanier, whom one may be tempted to make a hero for his blunt remarks, soon thereafter supported public funding for a new Astros ballpark. But he was right the first time, and that's one more time than most mayors are on this subject.

In 1997, Bud Adams moved Houston's football team to Memphis and then Nashville, where they would become the Tennessee Titans. Music City rewarded Adams with a $292 million stadium, half of which was paid for by a countywide water tax. As economist Raymond J. Keating noted in "The NFL Oilers: A Case Study in Corporate Welfare," Bud Adams started out as a gutsy entrepreneur, one of the AFL's charter members in 1960, only to become a flouncing corporate welfare queen.[112] He moved into the Astrodome, whose sordid history we have already examined, and in the late 1980s Adams bloviated and threatened to move to Jacksonville unless Houston's taxpayers forked over $67 million for improvements to the Astrodome — which they did, through boosts in the hotel and property taxes.

By the early 1990s, Adams was pressing Houston to build him a new Astrodome. That's when Mayor Lanier showed a spine not often seen in American mayors. He told Adams not to take a hike but to quit asking for a bailout. He denounced the idea of "taking Joe Sixpack's money and putting it into supporting a stadium...for owners with $100 million assets and players making $1 million-plus on salary." The *Houston Post,* exhibiting a refusal to shill for sports corporate welfare that is rare among American newspapers, editorialized, "If it is a good deal, private enterprise will do it."[113]

Nashville's civic leaders didn't see it that way. They put together an offer that a man like Adams, accustomed to corporate welfare, couldn't refuse. The stadium was to be funded almost entirely by various levels of government: $144 million from Nashville, $55 million in construction bonds and $12 million in road improvements from the state of Tennessee, and $70 million in personal seat licenses (to be guaranteed by Nashville taxpayers). Angry taxpayers forced a public referendum, wherein they were outspent 16–1 by the "Yes for Nashville" campaign. Despite this lopsided spending imbalance, the stadium package passed with only 59% of the vote. But that was enough. Tennessee got the Oilers, who metamorphosed into the Titans and today play in LP Field (which, like other stadiums whose naming rights are for sale, has gone through names like the Washington Redskins go through coaches). At least Nashville's mayor, Phil Bredesen, was frank enough not to defend this subsidy on economic grounds: "I can't justify building a football stadium on direct economic impact. The professors who make a living pooh-poohing that are right. But there are a lot of intangible benefits that make it more than easy to do."[114]

Back in Houston, Mayor Lanier suggested a new name for the former residents of the Astrodome: "the Nashville Welfare Mothers."[115]

America's Team?

In Dallas, where nothing is ever understated, the Cowboys of the NFL opened their $1.12 billion domed stadium with a retractable roof in 2009 to derisive jeers from connoisseurs of tasteful architecture and screams of "ouch!" from pinched taxpayers. The three million square foot stadium, which is three times as big as the retractable-roofless Texas Stadium which it replaced, is a gaudy edifice befitting a man like Cowboys owner Jerry Jones. Complete with "gargantuan" 72 × 160 foot video screens — the world's largest high-definition video screens, don't you know — and 315 luxury suites, Cowboys Stadium is a monument to a team which *Forbes* has ranked as the most valuable team in professional sports ($1.6 billion).[116]

But just because the Dallas Cowboys are a goldmine, a veritable gusher of football wealth, doesn't mean that Jerry Jones spent his own dime to furnish its home. The city of Arlington, which seemingly likes to be beggared by rich owners — witness the Rangers — chipped in $325 million in construction costs, that sum raised through a half-cent sales-tax increase, a 2% hotel occupancy tax increase, and a 5% car rental tax increase. Hotels and cars: the favorite target of the stadium taxers these days. Voters approved the tax increases — which foes dubbed "the Jones tax" — in 2004. The league also kicked in $150 million.

Before the Arlington deal, the city of Dallas almost lassoed its Cowboys back within the city limits. When Jerry Jones went looking for a new home in anticipation of the 2008 expiration of his lease at Texas Stadium in Irving, he dangled in front of Dallas's Staubach-spangled eyes the prospect of returning to the Fair Park neighborhood in which the Cowboys played (in the Cotton Bowl) until their move to suburban Irving in 1971. There was a catch, of course, and by catch Jerry didn't mean something that Bob Hayes or Drew Pearson pulled down. No, he offered in 2004 to return the Cowboys to Dallas if the Dallas County commissioners would put up $425 million of what was then an estimated $650 million project. It was a case of "I'll be happy to move into your city — if you buy me a house."

Even the most brazen beggar usually lacks the chutzpah to ask for that kind of give-a-way, but team owners are a rare breed. Despite gushing by Dallas Mayor Laura Miller — "This is the greatest thing that could ever happen to South Dallas" — the commissioners of Dallas County balked at the price tag.[117] And then Arlington stepped in.

(The city of Dallas was removed from the competition in part because of the terms of a previous subsidy. It was prohibited from expending any new monies on a stadium until it had retired the debt on the $420 million American Airlines Center, which the city had built in 1999–2001 to house the Dallas Mavericks of the NBA and the Dallas Stars of the NHL. The center was funded in part by hotel and rental car taxes; any further tax increases to build a football stadium would, by terms of the contract, fall only on non-city residents of Dallas County — and you can imagine how well that went over with the people of Dallas County. Predictably, advocates of the hockey–basketball arena asked voters, "Do we want our community to be considered Major League or Minor League?")[118]

Barely two months after the Dallas media was trumpeting the potential renaissance of a football-palace-hosting South Dallas, Arlington was negotiating with the Cowboys for the high honor of taxing its citizens to make Jerry Jones richer. The city council quickly and unanimously voted to put the cluster of tax increases — sales tax, car rental tax, hotel tax — on the November 2004 ballot.

Arlington had condemned and then stolen land in the early 1990s to build a parking lot for the Texas Rangers, so the thought of eminent domain on behalf of the Cowboys sprang naturally to mind. (Arlington had paid off the bonds by which the baseball park had been financed and was ready for new obligations, it seems.) The city's offer to the owners of 12.5 acres whom it was shaking down was $1 million; a civil court in Tarrant County readjusted the fair price to $5 million. So even those Arlingtonians who wouldn't mind having their homes taken and given to Jerry Jones had reason to fear the probable unfairness of any new "purchase" offers from the city of Arlington.[119]

Arlington's mayor, who was saddled with the unfortunate name of Robert Cluck, assured the people he ostensibly represented that eminent domain for the Cowboys would be a last resort, but few were buying that. Brenda Parker, a long-time resident of a trailer park in the targeted area, told the Fort Worth *Star-Telegram* that she feared what might happen should the "Jones tax" be passed. "You're uprooting family ties and history. I know it seems easy for us to move our home because it's on wheels," but those wheels had stayed put for 15 years, and why should she be forced to move just because Jerry Jones (estimated net worth: about $2 billion) wants her land?[120]

What Jerry Jones wants, Jerry Jones gets. And he did so with the acquiescence — the approval — of the people of Arlington. On November 2, 2004, they voted by 55–45% to tax themselves, or at least tax those who buy things, and those who rent cars or stay in hotels.

Mayor Cluck borrowed an oratorical trope from Martin Luther King Jr. for the occasion. "We had a dream," he told cheering supporters of the Jones tax. "We all had a dream, and that dream is now reality."

The dream was made reality in part by a Cowboys-sponsored $4.6 million advertising campaign that outspent the anti-giveaway Concerned Taxpayers of Arlington ($118,000) by about 40–1. The disparity in spending on television ads was 100–1. Even the Dallas Cowboys cheerleaders, subject of many an adolescent fantasy over the last three decades, got into the act, decorating stadium campaign events.

Among the weapons deployed by the Cowboys was a study commissioned by the City of Arlington and done by the Economic Research Associates that claimed — time to take a humongous grain of salt — that the new stadium would enrich the local economy by $238 million annually. Other economists scoffed at these numbers, but those other economists didn't have an almost $5 million advertising juggernaut proclaiming their words as gospel truth.[121]

To sweeten the pot, Jerry Jones threw in "a pledge to donate $16.5 million to youth sports."[122] Legislators assuaged local concerns over Cowboys-clogged traffic by leapfrogging Arlington transportation projects over those of other cities. The Cowboys and their neighbors-to-be, the Texas Rangers, drew up a plan for a town center along which, it was claimed, would occur the development which is always promised but seldom delivered by such projects. And of course the Cowboys legends of the past — Roger Staubach, Troy Aikman, and Emmitt Smith among them — took to the hustings, and who in the name of Tom Landry could refuse them? The stadium deal was a shoo-in.

But Cowboys Stadium was built on more than just taxpayer subsidies. Arlington would use its power of eminent domain to condemn over 150 homes and apartments and small businesses that stood in the way of Jerry Jones and his behemoths in blue. The property rights of the people whose land Arlington seized and gave to Jerry Jones were barely even afterthoughts in the national coverage of Cowboys Stadium. The stories of the displaced were judged to be nowhere near as compelling as the stories of Tony Romo's love life or the agonies of Jerry Jones as he watches the Cowboys, the erstwhile America's Team, lose clutch late-season games year after year.

But occasionally the local press paid some attention to the losers in this mismatch. For instance, the *Dallas Morning News* gave mention to Mrs. Maria Villareal, who raised a dozen children and stepchildren in a home on Vine Street in which she had lived for over 40 years. The Villareals were kicked out of their home in 2006, though perhaps it was a blessing that the 82-year-old Mrs. Villareal died just days before she was to have been moved out of her neighborhood.

"All my memories of growing up are from that house," son Troy Villareal told the *Morning News*. "Not only is that now gone, but my mother is gone, too." The hearse that carried Mrs. Villareal's body made a ceremonial drive down Vine Street, through the "once-familiar neighborhood that has become a ghost town."[123]

That ghost town is now Cowboys Stadium, and Jerry Jones had better hope that angry spirits don't deflect Cowboys passes and field goal attempts. Buildings have been haunted for far lesser reasons than eminent domain.

Arlington spent approximately $80 million in what was euphemistically called "land acquisition" but might more accurately have been dubbed "payment for stolen land." Numerous home and business owners took the city to court over its low-ball offers. One homeowner, offered $351,000 for a house and four acres of land, was eventually awarded $2.75 million, though most of her fellow litigants in the eminent domain cases lost their suits. The last resident to settle, Paul D. Jordan, told the *Dallas Morning News* that the area which the Cowboys demolished was "one of those rare neighborhoods that you don't often find that had a sense of community. I knew everyone around me on a first-name basis."

That sense of community counted for nothing in the calculus of the city and the courts. "We were paying more taxes than [the Cowboys] will ever pay," he said angrily. "That's such a lopsided, one-sided, bad deal for anybody but Jerry Jones."[124] As Arlingtonian Linda Lancaster said looking back on the land-grab, "All that we nice Christian folks in Arlington ended up doing is using eminent domain to take homes and businesses from our fellow citizens and agreeing to fork over millions in taxes that could have been used for other city services."[125]

Glenn Sodd, who represented two landowners in the earlier Texas Rangers eminent domain case and also served as attorney for some of the homeowners who objected to the theft of their land for Cowboys Stadium, said, "It's a misuse of the Texas Constitution and the U.S. Constitution. To say we in this community need this stadium is a gross mischaracterization. We might desire it. We might wish to have it. But no one's condemning land to build grocery stores."[126]

The Institute for Justice, a libertarian public-interest law firm, has worked diligently on eminent-domain reform in Texas and throughout the country. The IJ thought the Texas state legislature was going to enact real reform via an amendment to the Texas Bill of Rights in 2009, but the amendment was so watered down as to be perhaps worse than useless.[127] For all the lip service Texas politicians like to give to rugged individualism, it would seem that a preponderance of them are toothless pussycats purring meekly in the laps (or hip pockets) of the most unscrupulous developers.

What is especially galling is that Jones seems to accept public funds as his birthright. "I could have built this for $850 million," he told the *New York Times*, "and it would have been a fabulous place to play football. But this was such an opportunity for the 'wow' factor... That's why I spent the money."[128]

Unfortunately, he was playing with other peoples' money, but there is no penalty for that in today's National Football League.

The Super-Costly Superdome

Richard Dodge, a supporter of the St. Petersburg dome, enthused, "There is something about a dome that excites people; they get more bullish on themselves and where they live. It gives them a new and positive view of themselves. They react to issues and challenges in different ways. You can say it is the pride factor."[129]

A dome also costs scads of money.

Among the stadiums of the 1970s, the Superdome in New Orleans ($163 million to construct in 1971–1975, as compared to $22 million for Buffalo's outdoor Rich Stadium, built in 1973) was a classic monument to overspending.

The Louisiana Superdome Authority was created in 1966 by a three-to-one vote in a statewide proposition. The Authority was authorized to borrow $35 million to construct a facility to host major league sports contests, trade shows, concerts, and conventions. Bernard Levy, who served as executive director of the Louisiana Stadium and Exposition Commission, told *Newsweek* in 1974, "That $35 million was just a wild-hair guess. They had no site selected for the project. They had no architectural plans drawn. They were thinking of a 55,000-seat stadium. They hadn't included parking facilities."

That $35 million "wild-hair guess" looked more ridiculous as the months and years passed, until Governor John McKeithen remarked in 1971, "We're not going to let a couple of million dollars stop us."[130]

A couple of million? Try $128.5 million extra, as the final cost of the Superdome was $163.5 million.[131] There was fraud, there were kickbacks, and all the usual elements of a Louisiana political gumbo. But the Superdome was at the heart of the New Orleans urban redevelopment strategy, and no price was too much to bear.

The Superdome was a typically squalid product of Louisiana politics. As Robert K. Whelan and Alma H. Young write in "The Politics of Planning and Developing New Sports Facilities: The Case of Zephyrs Park and the New Orleans Arena," "Mayor Moon Landrieu and former governors

John McKeithen and Edwin W. Edwards were instrumental in getting the Superdome built. Patronage resulting from the construction of the Dome (e.g., concessions contracts) was used to reward their political supporters, including African–American politicians who were supporters of these white political leaders."[132]

The New Orleans Jazz of the NBA moved after four years in the dome (1975–1978), and the Crescent City never could attract a Major League Baseball team. The pathetic Saints of the NFL were the primary tenants, as were 35,000 displaced New Orleans residents in the wake of Hurricane Katrina in August 2005. Of course the Saints did have one magical season, capped by a victory in the Super Bowl in 2010. Thousands of delirious New Orleans fans probably would have said, the day after that game, that the Superdome had been worth it, cost overruns, corruption, and all. It certainly was worth it for the politicians and the contractors, at least. But they're not the ones who pay the bills.

And in an Even Worse Dome…

The Seattle Kingdome, which opened in 1976 with the expansion Seahawks of the NFL as the major tenant, was one of the worst of the publicly funded domes. Conceived in the 1960s and originally envisioned as a home for the Major League Baseball Seattle Pilots, the Kingdome was funded by $40 million raised through a bond issue approved by King County voters in 1968 as part of a broader development program with the bloodlessly progressive (or overtly sexual?) title "Forward Thrust." Earlier propositions had been defeated in 1960 and 1966, but the 1968 package included parks, community centers, and transportation initiatives totaling $334 million, of which the Kingdome consumed just over 10%.

Nevertheless, this turkey was jeered even in embryonic form. At the groundbreaking in 1972, Asian protesters, residents of the International District in which the dome would be built, booed loudly. And when the Kingdome was demolished less than 30 years later, on March 26, 2000, few mourned its destruction.

But many protested the way that its successor, Qwest Field, was built, for it had this in common with the unlamented Kingdome: both were constructed on the backs of taxpayers.

Seahawks owner Paul Allen of Microsoft — net worth: $12.7 billion — actually paid $4.2 million for the state of Washington to conduct a

June 1997 referendum on a $300 million tax package to build what became Qwest Field. The package included new state lottery games, an extension of the King County hotel tax, sales tax credits, and taxes on admissions and parking. Paul Allen would operate the facility with his company First & Goal.

The statewide vote was tantalizingly close: 51.1% yes, 48.9% no. And so in 2002, the Seahawks had a new home — bought, largely, by the people of Washington State for one of the richest men in the world.

Boston: Monuments to Free Enterprise

We find lessons in the potential of private arrangements and a relatively free market to provide playing fields for professional athletic teams in the strangest places. In Los Angeles, for instance, with the Dodgers. And in Boston, where the baseball temple of Fenway is a throwback in more ways than one. Not only is it a park redolent of the first golden age of baseball — a park so old that it predates the home run as a standard offensive weapon — but it is also a remnant of an age in which baseball teams, as businesses in the entertainment industry, built their own parks, without even bothering to ask the host municipality if it might have a spare million (or, today, hundred million) lying around.

True, in the late 1990s, Red Sox management pined for an eminent-domain-cleared new park for Fenway. Half of the more than $500 million cost was to be borne by taxpayers (for land acquisition, infrastructure, parking garage). This wretched and ahistorical plan cleared the legislature, but a series of missteps by Red Sox ownership queered the deal — a rare case of taxpayers winning. The baseball gods smiled.

But the football Patriots — born the Boston Patriots, and more broadly titled "New England" since 1971 — offer a lesson, too. The Patriots were a refreshing exception to the early 1970s trend toward 100% government-financed cookie-cutter stadiums.

Foxboro Stadium opened for business in the fall of 1971 to host the New England Patriots, who had been playing in Fenway Park — a great park for baseball, but maybe not so much for football. Located in Foxborough, Massachusetts, the stadium, which somehow lost its last three letters in a bout of simplified spelling, was privately built, a "bare bones edifice" privately financed by the sale of 400,000 shares of stock of the Stadium Realty Trust and constructed for just $6.7 million thanks to creative marketing.

Foxboro was a surprisingly late-era case of a stadium underwritten by frugality. Naming rights to the stadium were sold, first to Schaefer, the workingman's brewery, which also bought the exclusive right to sell beer onsite to thirsty Pats fans. (The sale of naming rights, if done by a private concern, is an exemplary way to help pay for a stadium — but not when cities and counties build the stadium and then permit the franchise to keep at least a portion of naming-rights fees.) The scoreboard at Foxboro was obtained in a trade for advertising, and the turf was "donated by a company trying to break into the stadium supply field." Conclude James Quirk and Rodney D. Fort in *Pay Dirt*, "The cost containment story of the stadium should be studied by anyone who thinks that the free enterprise system and private incentives can't work to keep costs down."[133] Foxboro, notes Dean V. Baim, had "the lowest cost per seat of any modern stadium."[134]

Unfortunately, the lesson of Foxboro was absorbed by virtually no one. Foxboro was one of five NFL stadiums built in 1970–1971; while its private/public financing ratio was 100/0, the other four structures that went up in that period (Riverfront in Cincinnati, Three Rivers Stadium in Pittsburgh, Texas Stadium in Dallas, and Veterans Stadium in Philadelphia) were perfect 0/100s — facilities constructed entirely with tax dollars.

Foxboro was derided as inadequate almost from the get-go. Luxury suites, plush locker rooms, state-of-the-art media facilities: the Patriots couldn't afford such extras, and the Commonwealth of Massachusetts — unlike so many Sun Belt states with reputations for fiscal conservatism, states whose residents jeer at the Bay State as "Taxachusetts" — refused to open the public purse to the Pats.

Belying their name, however, the Patriots had not been terribly loyal to Boston. In the 1990s, new owner Bob Kraft went city-shopping. He inveigled a deal out of the state of Connecticut to build a stadium with public monies for the Patriots in Hartford, but it came a cropper. So Kraft settled on keeping the team put, and he put up $325 million to build Gillette Stadium, a mostly privately financed stadium chockablock with luxury suites and club lounges, though the state did contribute $70 million for access roads and other infrastructure improvements. Schaefer Stadium, later known simply as Foxboro, was demolished in 2002 to make way for a parking lot for the new Gillette Stadium.

Gillette opened in 2002 to host the defending Super Bowl champs, the Patriots, and if it is a far sight from shabby old Schaefer/Foxboro, the stubborn and principled refusal of Massachusetts to subsidize the Patriots in the 1970s can be seen as an early act in the drama that led to Bob Kraft putting his own money — or at least some of it — where his team plays.

Taxing Cincinnati

In 1968, the voters of Hamilton County, Ohio, approved a bond issue to finance construction of a stadium on the Ohio River to replace aging Crosley Field. Crosley, then named Redland Field, had been constructed for $225,000 in 1912 after a series of fires had burned the park standing on this site three times in the previous dozen years.

Riverfront Stadium, which opened in 1970, cost taxpayers $45 million, with a portion (less than a third of that total) supplied by federal urban renewal monies.

Riverfront was the uber-cookie-cutter park which had so wowed Senator Marlow Cook in his hearings on a Federal Sports Commission. By the early 1990s, though, it was too old, too boring, too lacking in luxury suites, and too small (capacity was only 56,000) for Reds' owner Marge Schott and Bengals' owner Mike Brown. Riverfront had been home to the Big Red Machine, one of the great baseball dynasties of modern times, but it was hard to get sentimental about so sterile a venue. (It was so coldly 1970s that it was the first baseball park to cover its surface completely with artificial turf and to put metric distances on the outfield fences!) Schott made noises about pulling the Reds out and finding a suitable suburban location, perhaps even in Kentucky, and Brown put the Bengals' name into the hat whenever potential franchises on the move were discussed. The name *Baltimore* he frequently dropped.

City and county officials responded the way they usually do when major league teams threaten to move: they asked how they might be of service to the owners. By building us two new stadiums, replied the owners: a football-only field for the Bengals, and a baseball-only park for the Reds.

And so citizens in Hamilton County, Ohio, went to the polls to vote on a half-cent sales tax increase on March 19, 1996, in order to raise $540 million over 20 years to fund new playpens for the Cincinnati Reds (the ridiculously named — after an insurance company — Great American Ballpark) and Cincinnati Bengals (Paul Brown Stadium). Lest voters balk at shelling out enormous sums for rich owners and pampered players, the supporters of the sales tax increase threw in two sweeteners: funds for public schools, and a property tax rebate. They also launched a public-affairs offensive that overwhelmed the penurious opponents of the tax hike.

Political scientists Clyde Brown of Miami University and David M. Paul of the University of Illinois at Urbana-Champaign studied the Cincinnati case from an interest-group perspective. Sharing their findings with readers of the *Journal of Sport and Social Issues*, they took as their epigraph this lucid exposition of the problem by Vilfredo Pareto, writing in 1896:

> Those who hope to gain...will know no rest by day or night. Those who hope to gain...have agents everywhere, who descend in swarms on the electorate, urging the voters that sound and enlightened patriotism calls for the success of their modest proposal. In contrast, the individual who is threatened with losing [a small amount of money through taxation] — even if he is fully aware of what is afoot — will not for so small a thing forego a picnic in the country, or fall out with useful or congenial friends, or get on the wrong side of the mayor or the prefect! In these circumstances the outcome is not in doubt: the spoliators will win hands down.[135]

The spoliators swarmed over Hamilton County, Ohio, in 1995–1996.

Brown and Paul frame the interest-group analysis of such campaigns as follows: First, "those who stand to gain financially from the construction of new stadia (team owners, local corporate interests, downtown real estate interests, downtown restaurants and hotels, and financial institutions) will support the sales tax increase"[136]; second, "efforts will be made to distribute the costs as widely as possible among the public"; and third, "the more heterogeneous, broad-based anti-tax forces will have greater difficulty obtaining resources and getting organized" than does the pro-sales tax side.[137] And that is precisely what happened in Cincinnati. Concentrated benefits and diffuse costs usually add up to public subsidy.

Complicating matters — but only somewhat — was the obvious lack of public enthusiasm for such projects. A Louis Harris poll taken in spring 1995 found only 17% of Cincinnatians favored taxpayer subsidy of two new stadiums.[138]

But from the point of view of new stadium advocates, that poll only indicated that Cincinnatians needed to be *educated* into the right position. The commissioners of Hamilton County voted in July 1995 to impose a one-cent increase in the sales tax, which would boost the tax from 5.5% to 6.5%. (Ohio law required this to be accomplished by two separate half-cent increases.) A grass-roots "Citizens for Choice in Taxation" group arose, led by attorney Timothy Mara. Faced with the daunting task of collecting 27,000 signatures on each of two petitions in order to force a public vote on the sales tax increases, 700 volunteers collected 90,000 signatures on the pair of petitions. The people would have a choice.

That's when the public relations onslaught began. A front group calling itself "Citizens for the Future of Hamilton County," whose staffers included top Ohio advertising executives and political operatives, rolled into action.

In true boosterish fashion, they soon renamed themselves "Citizens for a Major League Future," implying, of course, that those who objected to the transfer of half a billion dollars of taxpayers' money into the pockets of such as Marge Schott and Mike Brown were bush leaguers.

The pro-stadium forces played the game by the most banal and hackneyed rules — but then those rules usually ensure success. The county commissioned an economic impact study from the University of Cincinnati which found — will wonders never cease? — a true cornucopia awaited the people of Hamilton County if only they would tax themselves. According to Brown and Paul, the impact study predicted a $1.13 billion "one-time economic benefit," an annual benefit of $296 million to the local economy, and the creation of almost 7,000 jobs.[139] Only a mossback afraid of the future and terrified of progress could possibly vote against the sales tax.

Still, the rent-seeking teams had a sneaking suspicion that the suckers weren't buying this story. Too many locals had attitudes similar to those of a pair of Cincinnati businessmen quoted in the *Wall Street Journal* during the campaign: "I feel the Bengals and Reds are engaging in extortion and blackmail"; and "As a conservative Republican, I have to ask: Why should government support private enterprise?" As the *Journal* noted, the battle along the banks of the Ohio River "pitted Cincinnati's corporate and political establishment against a scrappy antistadium group that is virtually tapped out for money...but is tapping into a rich vein of antitax, antiowner sentiment."[140]

So as 1996 dawned, the commissioners changed the shape of the tax plan. Instead of two half-cent increases it was now to be only one half-cent increase to fund the stadiums, with the Reds and Bengals supposedly chipping in, too; the other half-cent increase was now targeted to property tax relief and schools, which brought the teachers' union in line behind the plan. In addition, tracking polls commissioned by the pro-tax forces found that Cincinnatians were less than thrilled by the prospect of transferring their hard-earned money to rich owners, particularly owners as unsympathetic as Schott and Brown. So the Bengals pledged a relatively small amount ($25–$35 million) toward construction, which was played up by publicists as a significant contribution.

Mara, leader of the antitax forces, called it a "reverse Robin Hood scheme" under which sales taxes were taken "from the little guy...to transfer it to millionaires." This was indubitably so — but it didn't stop those "citizens" who desired a "major-league future" from pushing it. And the resources with which they pushed it were enormous. Brown and Paul note that whereas Citizens for Choice in Taxation "raised small contributions by passing the hat," the political campaign manager who was hired to run Citizens for a Major League Future said that he received this instruction from his bosses: "You tell us how much this is going to cost and we will raise it."[141]

The antitax group spent less than $30,000; its media blitz consisted of a single "30-second radio spot on one station." The pro-tax side spent over $1.1 million. Top contributors included the Bengals ($300,000 — it was the least they could do...); Procter & Gamble ($70,000), which is headquartered in Cincinnati; and the Northern Kentucky Chamber of Commerce ($35,000). "[T]hose who stood to gain the most from public funding of the sports stadia were prominent in financially supporting" the campaign, write Brown and Paul.[142] These included the Bengals, real-estate interests, developers, and banks. The Reds are an obvious omission from that list, though the unpredictable Marge Schott kicked in $41,000 after the election to pay off the entire debt of Citizens for a Major League Future. Still, Schott, who had a long history of feuding with the Bengals, made clear her indifference to the football team's fate, at one point telling the *Dayton Daily News*, "I'm sick of that stadium thing. I have one question: Why do the Cincinnati Bengals need a stadium for 10 games a year? It makes no sense."[143]

In a postmortem on the campaign, the director of Citizens for a Major League Future said that his group's aim was to "set the terms of the debate, shut them down at every opportunity, [and] marginalize the opposition." This was war conducted by other means.

Citizens for Choice in Taxation insisted that the Reds and Bengals were bluffing about moving; the grass roots leaders argued that professional sports were a profitable enough enterprise to attract private investment in new stadiums. Ohio Governor George Voinovich denied this, saying, "There's only one question going into the next century: Is Cincinnati going to be a major-league city? And whether you like it or not, you are not considered major league unless you have franchises."[144]

Anxiety over the possible loss of either or both teams was skillfully exploited by the Citizens for a Major League Future. Hamilton County Commissioner Bob Bedinghaus, author of the sales tax plan, played the dread bush-league card: "There is a whole bunch of...second-tier cities creeping up on our heels. Unless we continue to provide a viable, exciting community, we're going to wake up and wonder how Nashville, Charlotte, even Albuquerque outran us."[145] A TV commercial warned, "Dozens of cities would love to steal our major-league image. They want it, but we've got it. Let's keep it. Vote yes on Issue 1."[146] The Chamber of Commerce chimed in, according to Brown and Paul, that Cincinnati "was on the verge of becoming another Memphis."[147] This kind of rabid boosterism and race-for-the-top-ism would be funny were it not for the hundreds of millions of dollars at stake.

The people of Hamilton County supported the half-cent sales tax increase by a healthy margin of 61–39%, though they roundly rejected the other half of the package, the half-cent sales increase for education and (allegedly) property tax reduction, by 63–37%. All told, Hamilton Countians would pay via taxes 76% of the cost of the two parks, with the rest of the burden falling on the state of Ohio, the Bengals and Reds, and a small contribution from the corporate sector.

Riverfront was imploded in December 2002, to the cheers of onlookers. The Great American Ballpark, cost $290 million, went up on the riverfront in time for the 2003 Reds' season.

The Bengals had moved into their new field three years earlier. At $455 million, Paul Brown Stadium has been called the plushest and most expensive halfway house in America, a nod to the Bengals' reputation as a team whose members are frequently just this side of the county jail. The Bengals have long been among the league's bottom feeders, but by threatening to move they provoked the cowed and credulous city fathers and mothers of Cincinnati to give the team whatever it wanted — and more. Paul Brown Stadium may represent the single most subsidized football stadium in the NFL, and it was built for a chronically poor team with a disproportionate number of embarrassingly thuggish players.

A dozen years later, county auditor Dusty Rhodes called the stadium sales tax deal "a train wreck." The sales tax fund is insufficient to meet its obligations. As of 2010, the Bengals no longer had to pay rent to use the stadium. The Bengals also get to keep all revenue from naming rights, tickets, suites, and advertising. Even in the slanted world of pro sports, the Bengals "won a particularly lopsided lease," as the *New York Times* observed.[148] And they show no signs of rewriting the lease to cut the taxpayers of Hamilton County a break.

Milwaukeeans Need a Beer

Milwaukee, which threads in and out of the subsidized stadium story since the early 1950s move of the Boston Braves to Wisconsin's largest city, comes back into our picture in the 1990s.

After losing the Braves to Atlanta, Milwaukee greeted the Seattle Pilots after that forsaken team flew out of the Northwest in 1970 after a single season in Seattle. Bud Selig, who had been a minority owner of the Braves and sought to keep them in Milwaukee, was the owner of the relocated Pilots, which he named the Milwaukee Brewers. Major league baseball was back in Wisconsin, and all was well with the world — until Selig tired of Milwaukee County Stadium and threatened to move if the people of Milwaukee didn't buy him a new ballpark. The man who had denounced those responsible for the Braves move as "carpetbaggers" was perfectly willing to move — or threaten a move — in order to hold up the taxpayers. As is so often the case, it was a question of whose ox (or which city's taxpayers) was being gored.

Selig was likened to William Bartholomay, the man who moved the Braves, by local reporters with long memories. Milwaukee had been through this before — and lost. This time the city, or at least those who make its political decisions, were determined to keep the team. As Wisconsin Governor Tommy Thompson, putatively a conservative Republican, said, "Without the stadium, the Brewers leave. That's the bottom line."[149] The other bottom line, by now a rank cliché, was expressed by Robert Milbourne, executive director of the Greater Milwaukee Committee, who had remarked some time earlier, "without major league sports, Milwaukee would be like Des Moines."[150]

In 1994, Wisconsinites soundly thumped a proposed statewide sports lottery (64–36%) to pay for new Brewers stadium. But the stadium-seekers came back the next year and got what they wanted. The bottom line for taxpayers ended up at $310 million, raised by a 0.1% sales tax on purchases made in Milwaukee County and four surrounding counties. Brewers' ownership chipped in $90 million. Construction of what became Miller Park took an inordinately long period of time (1996–2001) and was cursed by accidents, collapses, and the deaths of three construction workers. The sales tax, which began in 1996, was originally slated to expire in 2014 but now seems likely to be extended through 2018 — or later.

Rocky Mountain High (Taxes)

In "Stealing Home: Political, Economic, and Media Power and a Publicly-Funded Baseball Stadium in Denver," published in the *Journal of Sport and Social Issues,* George H. Sage analyzed the successful 1990 campaign to impose a sales tax on metropolitan Denver, the proceeds of which were to fund construction of a park for the city's expansion Major League Baseball team.

Setting aside the question of whether or not this was a wise use of consumers' tax dollars, Sage examined the ways in which the city's establishment sought to "convince taxpayers that they should build a stadium for the use of a privately owned corporation," and he also asked why "citizens were not more active in questioning the public expenditure of money for stadium construction or the terms" of a very generous lease.[151]

Denver had been frustrated before in its quest for a major-league team. Colorado Rep. Tim Wirth (D) had used his position of influence in the U.S. House of Representatives to lobby for a team for his state. He and Florida Republican Connie Mack III, grandson of the half-century-long Philadelphia Athletics manager and a proponent of — surprise! — an expansion team in Florida, presided over the hopefully named Congressional Baseball Expansion Task Force. The threat, not so much implicit as promised, was that if Denver and Miami did not get expansion teams, Congress would take a good hard look at baseball's antitrust exemption.

So when in 1990 Major League Baseball announced that it would expand by two teams in 1993, Denver (and Miami) was thought to be near the front of the line. But the Mile High City needed a venue other than Mile High Stadium, where the NFL's Broncos had first dibs on scheduling.

The Colorado legislature, sensing that a team was there for the taking, created a Denver Metropolitan Major League Baseball Stadium District which was given the power to request of voters in a six-county area that a sales tax of one-tenth of 1% be levied in order to subsidize construction of a stadium to host Denver's prospective team. The taxpayer contribution was said to be $97 million of the $139 million projected cost. As usual, the cost was underestimated; Coors Field, as the park came to be called, cost $215 million, $168 million of which was raised through the sales tax.

A vote was scheduled for August 14, 1990. The big guns moved into place.

Coors Brewing, Oscar Mayer, McDonalds, Philip Morris: the list of corporate donors to the Baseball for Colorado Committee was impressive, and not all its patrons were even Colorado-based. The board of the Denver Metropolitan Major League Baseball Stadium District authorized an economic impact study by an accounting firm. The study claimed, to the surprise of absolutely no one, that a major league team in Denver would inaugurate an age of prosperity the likes of which the Rocky Mountains had never seen — or at least it would be good for $25 million in annual revenues for Denver and an "overall economic impact" of $93 million.[152] (When pressed about the accuracy of such predictions, a financial analyst at Touche Ross, which had prepared an earlier economic impact study for Denver, conceded, "This is clearly an inexact science.")[153]

There was not enough salt in the Mountain Time Zone with which to take such a study, but it was good enough for the *Rocky Mountain News*, Denver's biggest daily newspaper, which acted as an "unabashed cheerleader" for the sales tax measure. The paper's editorialists made polemical use of the hollow economic impact study: "[V]ote yes...we're willing to pay a few dollars a year in taxes for the pleasure of the game and the $25 million (at least) that will be pumped into the economy every year as a result of baseball's presence...Suffice to say that lots of money would flow into the city, and plenty of jobs would result, many of those appropriate for young people just starting their climb up the job ladder." (If one starts one's career climb up the job ladder selling peanuts on 80 summer nights a year, it will be an *awful* long climb to self-sufficiency.)

The *Rocky Mountain News*, notes George H. Sage, was not content with shilling for the sales tax. It also belittled opponents of the tax hike as "forces of caution and stagnation," "short-sighted," "envy-wracked," and "skeptics" who see the world through "dust-covered glasses."[154]

The campaign succeeded. Voters in four of the stadium district's six counties approved the measure, which won by an overall margin of 54–46%. Denver got its franchise and its stadium. The ownership group of what became the Colorado Rockies included the *Rocky Mountain News*, which raised "major ethical and legal questions," says Sage, for the *News* had "used the power of the press to move opinion in favor of a public subsidy for a private venture that it secretly wanted to acquire an ownership interest in."[155] The paper had, in effect, campaigned for a sales tax hike for its own benefit. Justice or karma, however, soon exacted its price: the *Rocky Mountain News* ceased publication in February 2009, victim of the general decline in newspaper readership. No sales tax boost would ride to its rescue this time.

The Rockies would play the 1993 and 1994 seasons at Mile High Stadium before moving to Coors Field (naming rights purchased by the local brewery) in 1995. Sage details the beyond-sweetheart lease the Rockies received from the stadium district. Not only did the lease hand over all concessions revenue to the team, as well as revenue from non-baseball events, it even gave them the profits from naming rights. This was a gift outright, and in a striking example of the revolving door slamming shut on taxpayers' faces, the board chairman who negotiated the "deal" on behalf of taxpayers was soon named executive vice president of the Rockies.[156]

Disgust with the whole deal was widespread. As one observer commented, "the community voted new sales taxes to finance the $215,000,000 Coors Field, while they voted down a $32,000,000 bond issue for schools. It does not make sense."[157]

But given the wretched state of public education in the cities, can you blame them? Is there any reason to believe that $32 million would have done much more than find its way into bureaucratic pockets?

Charge!

The San Diego Chargers have played in the unlovely named Qualcomm Stadium, nee San Diego Stadium, since 1967, which in the NFL is the equivalent of playing against the Cro-Magnon II on the fields of prehistoric Europe. San Diego Stadium first hosted the Chargers of the AFL. In 1969, the expansion baseball Padres moved in. Built by the City of San Diego and paid for by a $27 million bond issue, San Diego Stadium was just another cookie-cutter multipurpose stadium subsidized by taxpayers. It was built in the "Brutalist" architectural style, a term which accurately conveys its charm. It was renamed Jack Murphy Stadium in 1980, honoring a San Diego sportswriter who had lobbied for its construction, but in 1997 Qualcomm, the San Diego-based wireless telecommunications company, bought naming rights, and poor Jack Murphy was dumped into the trash heap of discarded names. Qualcomm paid $18 million for the naming rights; another $60 million was raised through a bond issue as part of a $78 million renovation in 1997. (This was done over the opposition of a "ragtag band of libertarians and anti-tax activists," in the words of *USA Today*.)[158]

Qualcomm Stadium is owned by the city of San Diego. Mayor Jerry Sanders of San Diego has boasted that Qualcomm has the biggest parking lot west of the Mississippi.[159] It is not clear if any other mayor in America cared enough to contest that claim.

Since 2002, the team has been agitating loudly for a new stadium. They have found a site near Petco Park, home of the San Diego Padres, and an engineering company has designed an inflatable roof stadium that would seat 62,000. As Mark Fabiani, a Chargers official, puts it, "No one in California expects to see a roof on a stadium because our weather is so great here." True enough. He explains, "It really turns a football stadium, which isn't used very much, into a multi-use facility — particularly in downtown San Diego, where it's very close to the convention center."[160]

The hitch: while the estimated cost of a new Chargers Stadium is $800 million, the Chargers have offered to put up only $200 million toward that princely sum. The NFL may lend the Chargers $100–200 million more. The rest will be the responsibility of the people of San Diego, and the people of San Diego have not exactly been charging the barricades demanding that they be taxed for this coliseum. Informal polls have showed widespread opposition to building the Chargers a pleasure dome.

Rent-seeking (and rent-evading!) football officials, politicians hankering to claim credit for a sexy "job-creating" project, and "ragtag" citizens: the elements are there for a Battle of San Diego in the near future.

"The Most Important Event"

The case of Jacksonville, Florida, is particularly poignant, or perhaps pathetic. So desperate was the city for a National Football League team that when the eccentric Baltimore Colts owner Robert Irsay flew by helicopter into Jacksonville's Gator Bowl on an August night in 1979, 50,000 football fans chanted, "We want the Colts" in what then-Mayor Jake Godbold called "the single most important event in the modern history of Jacksonville."[161]

Really, Mr. Mayor? Chanting and bowing and salaaming before a rich man touring the country trolling for handouts is the *single most important event* in your city's history? Irsay chose Indianapolis instead, and Jacksonville was also snubbed by Bud Adams and the Houston Oilers, who in 1987 rebuffed the city even as it promised 10 years of sold-out games.

Jacksonville campaigned for a team with an almost frightening avidity. The city pledged to spend $120 million on the new team — it even, noted the *Wall Street Journal* with some incredulity, promised to buy furniture for the team's office. But then as Thomas F. Petway III, a member of the Jaguars ownership group, said, "It's going to take us into the 21st century. It's going to make us a first-tier city."[162]

The Jaguars began play in 1995. And yes, the city did make it into the twenty-first century, along with every other city in America. But the pipe dream of one Jacksonville booster that "TV sets in…Cleveland, Pittsburgh, and Cincinnati will be showing Jacksonville's sunshine in the dead of winter, luring future tourists" was so much smoke, empirically speaking.[163]

The city fathers of Jacksonville, while campaigning for the team that eventually became the Jaguars and plays to the least-filled stadium in the NFL, left reason at the curb and claimed (backed by a University of North Florida study) that the Jags would bring in $130 million of new spending and 3,000 new jobs.[164] Robert A. Baade and Allen R. Sanderson suggest that the actual effect was perhaps "one-tenth as large as that estimated by boosters in Jacksonville."[165]

And that was with a new publicly financed stadium in Jacksonville for a new team. Knocking down "old" stadiums — that is, concrete bowls built perhaps 30 years ago — and replacing them with state-of-the-art luxury-box-filled retro parks complete with sushi bars and personal seat licenses gives the local construction industry a brief shot in the arm but does not appreciably add to the economy activity of the region. But yes, even the most jaundiced observer has to admit: TV sets in Cincinnati are tuned once a year to a football game played in Jacksonville.

New York, New York

Cowboys Stadium wasn't the only megabucks football field to open for play in the NFC East Division. The New York Giants and Jets christened their New Meadowlands Stadium in 2010, the season after Dallas unveiled its monument to excess. Built for $1.6 billion, thereby eclipsing Cowboys Stadium as the priciest home of football in America, this became the most expensive privately funded stadium in America. Or partly privately funded. For the stadium is built upon property belonging to the New Jersey Sports and Exhibition Authority, to which the teams make payments in lieu of taxes. Moreover, as the authority's former chief executive, George Zoffinger, pointed out to the *New York Times*, "New Jersey taxpayers are on the hook for about $400 million in road improvements, a new rail link from Secaucus and more than $100 million to retire the debt on the old stadium," in the *Times*' paraphrase.[166]

Unfortunately for the good folk of the Garden State, the unlovely stadium that was demolished to make way for the unimaginatively named New Meadowlands Stadium retains a rather urgent claim on the public purse. As Ken Belson discovered in an investigation for the *New York Times*, even as the Giants and Jets kicked off in September 2010 the erstwhile Giants Stadium still "carried about $110 million in debt, or nearly $13 for every New Jersey resident." (If it's any consolation — which it surely is not — the sports fans and non-sports fans alike of King County, Washington, still owed $80 million on the unlamented Kingdome, which was sent to that great dome in the sky in 2000. And the good people of Indianapolis won't finish paying off the $61 million in debt remaining on the also-gone Hoosier Dome till 2021.)[167]

As Belson noted, Giants Stadium was not supposed to cost New Jerseyans a cent. The adjacent Meadowlands Racetrack, which opened contemporaneously with the football stadium, was going to turn such a profit raking in sucker bets on the ponies that state taxpayers wouldn't have to pony up anything for the honor of hosting the Giants. True, the Giants refused to discard the name "New York" in favor of "New Jersey," but the Jerseyites could take solace from the baseball experience of Anaheim. The name thing was but a minor slight.

But the best-laid plans of New Jersey politicians often fall off the table. An array of newly legalized gambling opportunities — Off-Track Betting in New York, Atlantic City's casinos, and state lotteries everywhere you looked — ate into the profits of the Meadowlands Racetrack. The ponies didn't pay for Giants Stadium. Biped New Jerseyans would.

And to top it all off, Gregg Easterbrook of ESPN.com reported just three weeks into the 2010 season that "Season-ticket holders…don't merely dislike, but despise the New Meadowlands Stadium, which has none of the character of the previous Giants-Jets venue, positions spectators far from the action and doesn't generate that shaking feeling in the stands." To have none of the character of the previous Giants-Jets venue, which was about as characterless a stadium as they come, seems an almost metaphysically impossible feat, but with $1.6 billion all things are possible.[168]

A Twin Killing

In the 1950s, Horace Stoneham bought land in western Minneapolis with an eye toward moving his Giants halfway across the country, but when Walter O'Malley took the Dodgers all the way to the West Coast Stoneham followed suit. Still, as Major League Baseball went westward, Minneapolis made its bid for a team. The American League told Minneapolis that it needed a major league ballpark before it would be considered for a team, whether expansion or relocation, so the city issued revenue bonds (purchased by local businesses and citizens as a civic act) in the amount of $4.5 million to acquire a site in nearby Bloomington and build Metropolitan Stadium, which opened in 1956 with AAA tenant the Minneapolis Millers.

Minnesota lured Calvin Griffith and his Washington Senators in 1961. The expansion Minnesota Vikings of the National Football League also began play in 1961, also at Metropolitan Stadium.

The Met, as it was known, was a much praised baseball venue but the Vikings were unhappy with it. The Met was the smallest stadium in the NFL, with a capacity of under 50,000, though it had a great frozen tundra atmosphere. Opinion polls suggested that most Minnesotans preferred to renovate the Met to accommodate the Vikings, but the team demanded a new field. So the atrocity known as the Hubert Humphrey Metrodome was born — financed by $55 million in revenue bonds by the state and opened in 1982. The frozen tundra feel was utterly lost in the sterile Metrodome, into which the Twins also moved. (The publisher of the Minneapolis *Star-Tribune* was a major backer of the Metrodome, in typical newspaper publisher fashion.)

Both Minnesota teams tired quickly of the Metrodome. They wanted new stadiums or they wanted out — maybe even out of the Land of a Thousand Lakes and in the case of the Vikings, down to the land of the Alamo.

Carl Pohland, billionaire Twins owner and the second-richest man in Minnesota, pushed unsuccessfully for a new taxpayer-subsidized stadium in 1996–1997.[169] He was roasted as "Po' Lad" by, among other anti-stadium forces, Paul Moberg, founder of GAGME (Grass-Roots Against Government-Mandated Entertainment), whose efforts included a pamphlet depicting Pohland wearing a diaper and crying "My name is Carl. Please help. Please, won't you give to Billion-Aid? Tonight, somewhere in this cold marketplace, a little billionaire boy cries out for help."[170]

The economic impact propaganda for a new stadium for the Minnesota Twins in the 1990s was so over the top that University of Chicago economist Allen Sanderson said that if the subsidy money were simply "dropped out of a helicopter over the Twin Cities, you would probably create eight to ten times as many jobs" as the boosters claimed to create.[171]

In 1998, voters in the Greensboro, North Carolina, region soundly rejected a package of restaurant and ticket taxes that were to have financed construction of a baseball stadium to lure the Twins, who would have changed into the Carolina Twins. The anti-freeloader forces were outspent by an incredible $716,000–$26,000 — yet they won.[172]

The Twins finally got their new ballpark, and they didn't have to go south to do it. Target Field in downtown Minneapolis, which opened for the 2010 season, ran up a bill of $545 million, of which $152 million was borne by the Twins. The almost $400 million public share is being funded by a 0.15 sales tax increase in Hennepin County.

The Vikings have not been so lucky. In 2010, during the debacle of quarterback Brett Favre's final season, snow collapsed the Metrodome roof, forcing the Vikings to play their first outdoor game in 29 years, at the University of Minnesota's home field. The team has asked the state legislature for a new stadium, without a roof. Vikings vice president Lester Bangley said in January 2011, "A roof does not provide any benefit to the Vikings. It also costs a couple hundred million dollars more in capital costs, in addition to the operating costs that are much higher for a covered facility."[173]

The legislature, seeing in domes not sterile football games but boat and RV shows, concerts, and the like, counteroffered with a dome, insisting that the Vikings, like the Twins, pay one-third of the cost. Lurking in the background is the threat of a move to Los Angeles, which Vikings owner Zygi Wilf is holding over fans' heads.[174] The Vikings, we may be sure, will get their stadium. The only question is how much the taxpayers of the Gopher State will be on the hook for.

Minnesota's Real Love…and Canada's

The Minnesota North Stars, who joined the National Hockey League in 1967, were among the best of the first wave of "new" NHL franchises who joined the original six teams of the modern league. Until that is, 1993, when owner Norman Green, who had bought the team in 1990 and was generally despised for ruining a once-proud franchise, took them to the hockey cold-bed of Dallas, where they played in the city of Dallas's $27 million Reunion Arena.

Minnesota did rejoin the league in 2000, thanks to a $130 million Xcel Energy Center arena for which the taxpayers of Minnesota kicked in $65 million and the city of St. Paul contributed $65 million in sales tax revenue bonding.

While hockey and basketball arenas are somewhat beyond the football stadium/baseball park focus of this book, Minnesota's departure for Dallas was emblematic of a strange and ongoing period in professional hockey: the leaving of northern hockey hotbeds for Southern cities with mild climes, no ice outside the freezer, and very open wallets.

The National Hockey League recklessly pursued a Sunbelt relocation strategy in the 1990s, mostly under Commissioner Gary Bettman, a New York attorney. Bettman took over a sport whose historic heart and soul had been in Canada and the northern tier of the USA and, lured in part by subsidized arenas and in part by a pipe dream of making hockey a televised spectator sport on a par with, if not football, at least baseball and basketball, Bettman oversaw the league's expansion into decidedly non-hockey cities. In a single decade, the league put teams in Nashville, Phoenix, San Jose, Dallas, Miami, Anaheim, Atlanta, Tampa Bay, and Raleigh while abandoning such cold-weather markets as Winnipeg, Quebec City, Minnesota, and Hartford. Some were new franchises and others were relocated existing franchises (most notoriously, the beloved Winnipeg Jets of the Canadian plains moving to Phoenix to become the unloved Coyotes). The plan was to locate teams in growing cities whose governments would build them state-of-the-art arenas. Which these cities did.

For instance, Nashville, not content with being the capital of country music and a storied city echoing with the chords of Hank Williams, Johnny Cash, and the Grand Ole Opry, decided that it needed a major league sports team — pre-Titans — to really break into the bigs. In 1995, the city of Nashville offered $20 million in taxpayers' money to any NHL team that would ditch its current host and move to Music City. It also offered a sweet lease deal at a new arena the city was building on spec. So impressed was the league that it awarded Nashville an expansion franchise in 1997. The team has since been in the lower echelon of the league, attendance-wise, and seriously considered a move to Hamilton, Ontario, before it was squelched by the NHL.

To the southwest, the taxpayers of Glendale, Arizona, supplied $180 million of the $220 million cost to build the Glendale Arena, now the Jobing.com Arena, for the Phoenix Coyotes, who took the ice in 2003. The franchise left Winnipeg when that city hesitated when faced with owner Barry Shenkarow's demand for a new, luxury-suite laden arena paid for by the people of Winnipeg. (Winnipeg returned to the NHL in 2011–2012 as the new home of the Atlanta Thrashers.)

The league's Sunbelt expansion has been "an unmitigated economic failure," according to economists interviewed by *Toronto Star* columnist Robert Cribb.[175] The Phoenix Coyotes have been in and out of bankruptcy court. Yet to the teeth-grinding frustration of Canadian hockey fans, the Sunbelt teams — whose fans often don't know the difference between icing and offside — have often done well in the standings. Tampa Bay, Carolina, and Anaheim won consecutive Stanley Cups in the mid-2000s.

In the 2010 season, as reported by James Mirtle of the *Globe and Mail* of Toronto, "The majority of the NHL's biggest attendance drops this season came in warm weather cities, with seven of the 10 'sunbelt' franchise among the 12 teams with a 1.6 per cent or more dip."[176] The Los Angeles Kings, who came into the league with the 1967 expansion and thus are not considered part of the Sun Belt gang, and the San Jose Sharks, who entered in 1991, before the Bettman era, are exceptions to the middling attendance story of the Southern and Southwestern American teams.

A lucrative, or even reasonable, television contract with one of the major networks — or even with ESPN, TBS, or a first-tier cable network — is as remote from the NHL as Thunder Bay, Ontario, is from the Tampa Bay Lightning. Even Phoenix owner Jerry Moyes admitted in 2009, "Hockey will not work in the South. Mr. Bettman's plan is not working out."[177]

But the Coyotes skate on — on thin ice, to be sure, paid for by people who think Bobby Orr is a rowing machine.

Follow the Bouncing Ball

As for professional basketball, "The NBA has had a much more tumultuous history than that of organized baseball," write Quirk and Fort.[178] There has been consistent movement of franchises throughout the last half-century-plus of the league's existence: four teams moved in the 1950s, five in the 1960s, six in the 1970s, two in the 1980s, none in the 1990s, and three in the first decade of the twenty-first century.

The Buffalo Braves of the National Basketball Association are an illustrative example. The team came into the league in 1970, playing at Buffalo Memorial Auditorium, a $2.7 million WPA project of 1939–1940. The Braves enjoyed moderate success on the floor and drew decent crowds — until John Y. Brown came to town.

Brown, son of a prominent Kentucky family, had, with several other investors, purchased Kentucky Fried Chicken from Colonel Harlan Sanders in 1964. As president of the company, Brown built Kentucky Fried Chicken into one of the most recognizable fast-food franchises in America. When he tired of chicken, Brown went into hot dogs, buying the Lum's chain. In his spare time, Brown married Miss America 1971, Phyllis George, who served as eye candy on CBS's *NFL Today* pregame show.

Upon purchasing a half-interest and then full ownership of the Braves from Paul Snyder in 1976–1977, Brown disposed of two future NBA Hall of Fame players, Bob McAdoo and Moses Malone. The Braves spiraled downward, and his demands for more-than-favorable terms of rent from the city grew stronger. Buffalo *Courier-Express* columnist Phil Ranallo, on June 14, 1978, described Brown's attempt to shake down Buffalo taxpayers: "The concessions John Y. would like, I'd guess, would include an arena whose rent is $1 a year, a guarantee of a full house no matter how lousy the team plays, payment by the city of the portion of player salaries that exceed the NBA minimum — and free drinks and food for Brown."[179]

John Y. Brown's publicists never could explain why Buffalo taxpayers should subsidize his team. The city dug in, so he pulled out. Having stripped the team of its talent and taken them from the NBA's upper tier into its depths, Brown swapped franchises with Boston Celtics owner Irv Levin, who promptly moved the Buffalo team to San Diego, where they suffered for several years before moving up the road to Los Angeles. As the Los Angeles Clippers, the old Braves were for a quarter century the worst franchise in the NBA — possibly in all of sports. But John Y. Brown made out like a you-know-what. In 1979 he was elected governor of Kentucky, and while he and First Lady Phyllis filled the pages of the gossip columns, the governor's rumored presidential ambitions never took flight, in part because of his admitted fondness for frequent trips to Las Vegas.

But the city of Buffalo did not give in to Brown. And the Braves left town.

Other cities did give in to demanding owners, often quite willingly. San Antonio, for one, saw the NBA as the pathway to "big league" status. In 1977, Mayor Lila Cockrell defended a taxpayer-subsidized $3.7 million improvement to the Convention Center, where the Spurs, who had just been admitted to the NBA, played, by saying, "We have a great opportunity through the gaining of national stature on the sports scene. This will help attract industry and assist our economy."[180]

The problem is, once you pay for them to play, further payments will come due. And you have to pay or they won't play. San Antonio built the Alamodome for the Spurs in 1993 (cost: $186 million, raised through a 0.5% city sales tax) and the SBC Center (later the AT&T Center) for the team in 2002 at a cost of $186 million, which was raised by the ever-popular hotel and car rental taxes. At this rate, a new arena is due any day now.

Not every city could be as fortunate as Milwaukee, which is a shining example of private philanthrophy building a home for an NBA team: the Bradley Center in Milwaukee, home of the NBA Bucks since 1988. The daughter and son-in-law of the late Milwaukee manufacturer Harry Lynde Bradley sponsored the construction of and then donated the Bradley Center to the State of Wisconsin: an extraordinary gift, though wouldn't you know it, the owner of the Bucks, Senator Herb Kohl (D-WI), has lately been agitating for a government-built arena with all the bells and whistles to replace the Bradley Center, which at 20-plus years of age is rapidly becoming a senior citizen in the world of professional sports. No good turn goes unpunished, it seems.

The NBA boasts other largely or wholly privately financed playing courts: the Palace at Auburn Hills, where the Detroit Pistons battle, and the Arco Arena in Sacramento are each financed by owners of the teams.

But in the NBA, as in other leagues, when the outs want in, city and county governments have to flash the cash.

Among the newest NBA arenas is the Oklahoma City Arena in Oklahoma City, which was built in 1999–2002 in the hope of attracting a team. Oklahoma City, for all its virtues and its capital cityhood and its population of over half a million people, was without a franchise in the NFL, MLB, NHL, and NBA. In the downcast eyes of boosters, therefore, it was "minor league." So in 1993, as part of a downtown improvement project, the Oklahoma City Arena was included in a city redevelopment scheme financed by a one-cent tax on sales within the city. The Arena, which opened in 2002, cost $89 million — quite a taxpayer investment for a building which housed only minor-league hockey and arena football teams.

But Oklahoma City, or at least those who tax its residents and design its public works, had bigger dreams. Hoop dreams, to be exact. When Hurricane Katrina forced the New Orleans Hornets of the National Basketball Association to find temporary lodging, Oklahoma City was quick to offer its arena. The Hornets made the arena their home in the 2005–2006 and 2006–2007 seasons, during which they drew very well. The Hornets returned to New Orleans, but Oklahoma City investors bought the troubled Seattle Supersonics and moved the team to OKC in 2008. To propitiate the Oklahoma City Thunder, as the team was now called — keeping with the audible nickname of its predecessor, though thunder is less than supersonic — voters approved the extension of the one-cent sales tax to fund $101 million in improvements to the arena and a $20 million practice facility. Predictably, the pro-sales tax forces called their effort the "Big League City" campaign. Oklahoma City was now "big league" — and the price tag showed it.

Seattle lost its team; Oklahoma City gained a team. As Coates and Humphreys note, this was a zero-sum game. They suggest that "a better approach to maximizing welfare might be for the NBA to expand the number of franchises so basketball fans across the country had their own local team to cheer for."[181] Oklahoma City would have a team, and so would Seattle. But public subsidies in some sense depend on restricting the supply of teams, thus forcing cities to bid against each other for the privilege of hosting one. The league much prefers these games of musical chairs.

Smelling a Rat in Brooklyn

And in the end, it all comes back to...Brooklyn.

One of the most brazen acts of government-in-the-service-of-sports is the ongoing power play by a developer named Bruce Ratner (net worth: $400 million) to turn the New Jersey Nets of the NBA into the Brooklyn Nets. It is not merely a question of relocating a perpetually mediocre franchise, albeit one that comes out of the red, white, and blue matrix of the storied American Basketball Association.

No, Mr. Ratner has something bigger in mind, and unfortunately, his big dreams impinge on the homelier designs of the people of the Prospect Heights neighborhood of Brooklyn — a neighborhood that is not blighted, is not a slum, is not a moonscape. It is, rather, a real neighborhood, with people and shops and businesses. And what Ratner is doing, writes George Will, is "an especially egregious example of today's eminent domain racket."[182]

His plan, which the politically plugged-in Ratner announced in late 2003, was for a 22-acre development which would comprise offices, high-rises, a hotel, more than 6,000 apartments, and a new arena for the New Jersey — presumably later-to-be-named Brooklyn — Nets. Being a modern real-estate developer in tune with the tenor of the times, Ratner did not waste his time negotiating with the numerous private owners who constituted most of the landholders on this site. (New York City's Metropolitan Transit Authority also had an eight-acre rail yard within the desired area, but connected operators like Ratner don't mind negotiating with government bodies. He bought the property for $100 million in a very questionable bidding process which saw a real-estate firm called Extell bid $150 million for those same eight acres — and lose the "competitive" bidding.)

Instead, as Damon W. Root of *Reason* magazine explains, Ratner "turned to the government — including his old Columbia law school pal Gov. George Pataki — for a bailout." New York's notorious Empire State Development Corporation (ESDC) acted as Ratner's muscleman. Contrary to the evidence available to the naked eye, the ESDC declared that the neighborhood was "blighted," which provided legal cover for eminent domain proceedings. The ESDC's ridiculous claim asserted that the presence, for instance, of "weeds" within these 22 acres constituted blight. The libertarian public-interest law group the Institute for Justice responded, "A finding of blight premised on underutilization or the presence of weeds in a yard is not a finding of *blight* — that is, not a finding that property is causing harm to surrounding properties — it is simply a finding that the government does not like what a property owner is doing with a particular piece of property." As Root notes, what property *doesn't* have a few weeds or is in some way "underutilized"? Are all such properties to be fair game for the seizers of eminent domain?

Moreover, to the extent that any land within Prospect Heights was under-utilized, it was largely because eminent domain had already seized a number of properties and "left them empty, thus *creating* much of the unsightly neglect" Ratner and the ESDC claimed to find.[183]

Ratner mused to *New York* magazine, "The Dodgers, the Dodgers, the Dodgers. That's nice nostalgia, but we have to get beyond that. In a metaphorical way, we have to get over the Dodgers. That's important. Because that talk represents the way Brooklyn used to be. And how one talks about the new Brooklyn is very important."[184]

Well, one difference between the "old" Brooklyn of the Dodgers and the "new" Brooklyn of Cleveland transplant Ratner is that the Dodgers paid for the land where Ebbets Field was built, while Ratner employed the police powers of government to forcibly seize land for the Nets' arena via eminent domain, which he paid for at a discounted price.

The Atlantic Yards project is estimated to cost $5 billion, with a cool billion coming from the taxpayers of New York City and State. The arena is designed by Frank Gehry, the superstar and superexpensive architect. As a Ratner associate gushed to *New York,* "What you'll see there is not typical Brooklyn brownstones. There's a skyline to this. This is the future. This is Robert Moses. This is Levittown."[185]

It is both troubling and revealing that Ratner's admirer compares him to Robert Moses, whose urban renewal and highway construction schemes displaced, by government fiat, some half a million people. As George Will writes archly, "The problem was, and is, that people live and work where Ratner wants to build."[186] In the tradition of Robert Moses, Ratner would use the power of the state to forcibly move those people — little people, people who lack the connections of Bruce Ratner. The *New York Times,* which has been friendly toward the project — just as the *Times* was a big booster of urban renewal in the 1950s and 1960s, dismissing those who had qualms about the federal bulldozer displacing millions of people in cities across America — admitted that Gehry's design "would radically alter the neighborhood." Actually, the *Times* has a direct stake in this project, which is why it has been so inactive in exposing Ratner's ripoff of the taxpayers. As the paper primly admitted in a distant paragraph far down one story, "Forest City [Ratner's development company]...is also the development partner in a new Midtown headquarters for the *New York Times*."[187] Well isn't that a coincidence!

Ratner has a way to go before equaling Moses as a people mover — but the principle is the same. Ratner's friends point out that he is a "lefty" politically, as though that somehow insulates him from charges that he is a greedy developer stealing people's property. In fact, "lefties" have been as avaricious in coveting the property of others as anyone, and if one dresses up this covetousness with the language of the "common good" it can even take on the appearance of nobility. After all, why should the little people — small shopkeepers, corner-store businessmen, owners of modest homes — stand in the way of the dreams of "lefty" Bruce Ratner, who only wants Brooklyn to prosper and enter its glorious future?

Ratner proposed, essentially, to use government power to change the nature of an entire neighborhood. Patti Hagan, a Prospect Heights resident, explained: "Everything he's doing is just homogenizing. He plops down this massive suburban-mall architecture in the middle of the neighborhood. It creates walls between the communities. Atlantic Center was *built* to be unfriendly...One of the nice things about Brooklyn always was that the Williamsburgh Savings Bank was the only tall building around. This was big-sky country. This is just one more step in the Manhattanization of Brooklyn." The brownstones of the neighborhood will be completely overshadowed, when not obliterated, by Ratner's gaudy development.

Hagan bemoaned that "corporate welfare" was enabling Ratner to destroy her cherished neighborhood.[188] But this was for the greater good, as eminent domain projects are always said to be. What matters the character of a neighborhood when riches in a developer's pocket are at stake?

Ratner's crew is nothing if not politically crafty. A cynic might say they paid off black community leaders and groups in order to forestall opposition to the project. The notorious ACORN (Association of Community Organizations for Reform Now) was one beneficiary. Was this out of a selfless concern for the well-being of Brooklyn's poor? Well, not exactly. Ratner's Atlantic Yards proposal included a plan by which half of the housing created would be "affordable" — an elastic definition which stretched upwards of

$2,000 rent per month — and it turns out that ACORN would be running the projects.[189]

Forest City called it a "community benefits agreement," though the beneficiaries tended not to be members of the community as much as organizations that purport, with varying degrees of accuracy, to represent interest groups within the community. Daniel Goldstein, founder of the anti-Atlantic Yards group Develop — Don't Destroy — Brooklyn, went so far as to say that the community groups receiving money from Ratner were obeying their "wealthy white masters."

Having black faces defend Atlantic Yards, a project whose principal players are lily-white, has political advantages. Rev. Clinton Miller of Brooklyn's Brown Memorial Baptist Church charges, "I think race was used from Day I to window-dress the project." Race, and Forest City's "contributions" to community groups, helped Ratner's group paint the opponents of Atlantic Yards as elitist yuppies. For instance, Bertha Lewis, executive director of ACORN, chalked the criticism of Atlantic Yards up to "white liberals."[190]

In November 2010, Ms. Lewis and ACORN, having been wracked by scandal after scandal encompassing matters from embezzlement to voter fraud to the infamous case — caught on camera — in which ACORN representatives advised undercover filmmakers posing as a pimp and a prostitute how to obtain a loan for a brothel, filed for Chap. 7 bankruptcy. Included in the filing was a $1 million loan from Ratner's Forest City. "Community support" doesn't come cheap. As the watchdog blog Atlantic Yards Report put it, even if Forest City "doesn't get its $1 million loan repaid, it still will have reaped far more in value from ACORN's support."[191] Bruce Ratner was so grateful that he was even photographed wearing an ACORN t-shirt. But urban blight, not mighty oaks, is going to grow from this ACORN.

Of course the Atlantic Yards promoters claimed all sorts of fiscal and even social benefits would flow from the project. When in 2009 New York City's Independent Budget Office released a report predicting that the city would lose $40 million over 30 years as a result of Atlantic Yards, a spokesman for the city's Economic Development Corporation was quick to jump on the report as "sloppy and contain[ing] numerous inaccuracies."[192]

In December 2009, Ratner sold half a billion dollars in tax-exempt bonds, underwritten by Goldman Sachs and Barclays Capital, to finance the $1 billion arena. (Another $113 million was promised by Mayor Bloomberg — not in his private capacity as a wealthy man but in his public capacity — and the rest was to be raised privately.) For $10 million a year over 20 years, Barclays bought the naming rights to the arena, which is now to bear the rather un-Brooklynish name of Barclays Center. The lack of an apostrophe is sure to bedevil Brooklynites, at least until the next name change.[193]

Ratner the real-estate developer was not immune to the effects of the downturn in the real-estate market, but when you have governments pumping hundreds of millions of dollars into your enterprises — well, you make do. Ratner sold the Nets to the controversial Russian tycoon Mikhail D. Prokhorov.

In April 2010, Daniel Goldstein, the last holdout against Forest City Ratner, as the developer took to calling itself, finally took the money and ran. Goldstein, who had founded Develop — Don't Destroy — Brooklyn, took $3 million from the Ratner organization to move out of his condominium in Prospect Heights.

Goldstein fought the project for as long as he could, but the wolf — in the ungainly shape of New York State, which in March 2010 had officially seized Mr. Goldstein's condo via eminent domain — was at the door. As documentary filmmaker Michael Galinsky told the *New York Times*, for Mr. Goldstein "to not have made some kind of agreement at this point would have been irresponsible. It was either move out in two weeks, or move out in two months and take what the state offered." What the state offered, it seems, was about one-sixth of what he got from Ratner.[194]

As this book went to press, the monstrous specter of Ratner's Atlantic Yards hovers over Brooklyn — and David is running out of stones to hurl at Goliath.

Notes

1. Glen Gendzel, "Competitive Boosterism: How Milwaukee Lost the Braves," *Business History Review*: 530.
2. Ibid.: 534.
3. Ibid.: 541, 543.
4. Ibid.: 548, 556.
5. Michael Benson, *Ballparks of North America: A Comprehensive Historical Encyclopedia of Baseball Grounds, Yards and Stadiums, 1845 to 1988*, p. 16.
6. Glen Gendzel, "Competitive Boosterism: How Milwaukee Lost the Braves," *Business History Review*: 554.
7. Ibid.: 563.
8. Michael Benson, *Ballparks of North America: A Comprehensive Historical Encyclopedia of Baseball Grounds, Yards and Stadiums, 1845 to 1988*, p. 8.
9. Bill Shaikin, "Economists: Stadiums Are Bad Investments," *Los Angeles Times*, May 11, 2005.

10. Ibid.
11. Neil deMause, "NYC baseball," www.fieldofschemes.com.
12. Neil deMause and Joanna Cagan, *Field of Schemes: How the Great Stadium Swindle Turns Public Money into Private Profit* (Lincoln: University of Nebraska Press, 2008), p. 249.
13. Richard Morin, "Public Financing Opposed, Poll Finds," *Washington Post*, November 9, 2004.
14. Dennis Coates and Brad R. Humphreys, "Caught Stealing: Debunking the Economic Case for D.C. Baseball," Cato Institute of Washington, DC, Briefing Paper 89, October 27, 2004, p. 2.
15. David Nakamura, "Coalition Vows to Fight Stadium; Education, Homelessness Are More Pressing Priorities, D.C. Group Says," *Washington Post*, October 6, 2004.
16. Adam Bernstein and Dana Hedgpeth, "Behind a Wall of Silence, Lerner has Built an Empire," *Washington Post*, May 3, 2006.
17. David Nakamura, "Coalition Vows to Fight Stadium; Education, Homelessness Are More Pressing Priorities, D.C. Group Says," *Washington Post*.
18. Neil deMause and Joanna Cagan, *Field of Schemes: How the Great Stadium Swindle Turns Public Money into Private Profit*, p. 260.
19. "An Open Letter to Mayor Anthony Williams and the DC City Council from 90 Economists on the Likely Impact of a Taxpayer-Financed Baseball Stadium in the District of Columbia," October 21, 2004.
20. Bill Myers, "Ballpark renaissance striking out in D.C.," Washington *Examiner*, June 1, 2010.
21. Dennis Coates and Brad R. Humphreys, "Caught Stealing: Debunking the Economic Case for D.C. Baseball," p. 3.
22. Ibid., p. 4.
23. Bill Myers, "Ballpark tax may be here to stay," Washington *Examiner*, June 1, 2010.
24. Bill Myers, "Ballpark renaissance striking out in D.C.," Washington *Examiner.*
25. www.andrewclem.com/Baseball/BallparkArlington.html.
26. Neil J. Sullivan, *The Diamond in the Bronx: Yankee Stadium and the Politics of New York* (New York: Oxford University Press, 2008), p. xiii.
27. Ibid.
28. Neil deMause and Joanna Cagan, *Field of Schemes: How the Great Stadium Swindle Turns Public Money into Private Profit*, p. 200.
29. "Mr. Steinbrenner's Threats," *New York Times*, July 18, 1993.
30. Quoted in Neil J. Sullivan, *The Diamond in the Bronx: Yankee Stadium and the Politics of New York*, p. 172.
31. "Ralph Nader Writes George Steinbrenner III," August 24, 2006, www.nader.org.
32. Jim Dwyer, "Breaking With History in the Bronx," *New York Times*, April 3, 2009.
33. "NYC baseball stadium subsidies: Do I hear $1.8B?" January 14, 2009, www.fieldofschemes.com.
34. Stephen J. Agostini, John M. Quigley, and Eugene Smolensky, "Stickball in San Francisco," in *Sports, Jobs & Taxes: The Economic Impact of Sports Teams and Stadiums*, edited

by Roger G. Noll and Andrew Zimbalist (Washington, D.C.: Brookings Institution, 1997), p. 387.

35. Quoted in Mayya M. Komisarchik and Aju J. Fenn, "Trends in Stadium and Arena Construction, 1995–2015," Colorado College Working Paper 2010–03, April 2010, unpaginated.
36. Tom Barnes and Robert Dvorchak, "Plan B approved: Play ball!" Pittsburgh *Post-Gazette*, July 10, 1998.
37. James Edward Miller, "The Dowager of 33rd Street: Memorial Stadium and the Politics of Big-Time Sports in Maryland, 1954–1991," *Maryland Historical Magazine*: 195.
38. Ibid.: 197.
39. Charles C. Euchner, *Playing the Field: Why Sports Teams Move and Cities Fight to Keep Them*, p. 122.
40. Bruce W. Hamilton and Peter Kahn, "Baltimore's Camden Yards Ballparks," in *Sports, Jobs & Taxes: The Economic Impact of Sports Teams and Stadiums*, p. 246.
41. Ibid., p. 253.
42. Dennis Coates and Brad R. Humphreys, "Professional Sports Facilities, Franchises and Urban Economic Development," University of Maryland-Baltimore County Working Paper 03-103, undated, p. 3.
43. Drake Bennett, "Ballpark figures," *Boston Globe*, March 19, 2006.
44. Jacob V. Lamar Jr. and Don Winbush, "India-no-place No More," *Time*, June 11, 1984.
45. Quoted in John Crompton, "Beyond Economic Impact: An Alternative Rationale for the Public Subsidy of Major League Sports Facilities," *Journal of Sport Management*, Vol. 18, No. 1 (January 2004): 45.
46. Mark S. Rosentraub, *Major League Losers: The Real Cost of Sports and Who's Paying for It* (New York: Basic Books, 1997), p. 149.
47. Mark S. Rosentraub, David Swindell, Michael Przybylski, and Daniel R. Mullins, "Sport and Downtown Development Strategy: If You Build It, Will Jobs Come?" *Journal of Urban Affairs*, Vol. 16, No. 3 (1994): 225.
48. Ibid.: 228.
49. Ibid.: 233.
50. Mark S. Rosentraub, *Major League Losers: The Real Cost of Sports and Who's Paying for It*, pp. 236–37.
51. Ibid., p. 238.
52. Mark S. Rosentraub, David Swindell, Michael Przybylski, and Daniel R. Mullins, "Sport and Downtown Development Strategy: If You Build It, Will Jobs Come?" *Journal of Urban Affairs*: 237.
53. Ibid.: 221.
54. Bob Andelman, *Stadium for Rent: Tampa Bay's Quest for Major League Baseball* (Jefferson, NC: McFarland & Co., 1993), p. 264.
55. Neil deMause and Joanna Cagan, *Field of Schemes: How the Great Stadium Swindle Turns Public Money into Private Profit*, p. 175.
56. "AT&T Park," www.ballparks.com.
57. *Sports, Jobs & Taxes: The Economic Impact of Sports Teams and Stadiums*, p. 1.

58. Neil deMause and Joanna Cagan, *Field of Schemes: How the Great Stadium Swindle Turns Public Money into Private Profit,* p. 106.
59. Charles C. Euchner, *Playing the Field: Why Sports Teams Move and Cities Fight to Keep Them,* p. 93.
60. Katherine C. Leone, "No Team, No Peace: Franchise Free Agency in the National Football League," *Columbia Law Review,* Vol. 97, No. 2 (March 1997): 500.
61. Kenneth Reich, "Pro Teams: A Big Value for Small Cities," *Los Angeles Times,* December 3, 1989.
62. Chris Metinko, "Could a new football stadium be in Oakland's future?" San Jose *Mercury News,* February 12, 2010.
63. Neil deMause and Joanna Cagan, *Field of Schemes: How the Great Stadium Swindle Turns Public Money into Private Profit,* p. 8.
64. Ibid., pp. 15–16.
65. Veronica Z. Kalich, "A Public Choice Perspective on the Subsidization of Private Industry: A Case Study of Three Cities and Three Stadiums," *Journal of Urban Affairs,* Vol. 20, No. 2 (1998): 213.
66. Katherine C. Leone, "No Team, No Peace: Franchise Free Agency in the National Football League," *Columbia Law Review*: 486–87.
67. Quoted in Andrew Zimbalist, "The Economics of Stadiums, Teams and Cities," in *The Economics and Politics of Sports Facilities,* p. 57.
68. George Will, "Modell Sacks Maryland," *Newsweek,* January 22, 1996.
69. Dennis Zimmerman, "Subsidizing Stadiums," in *Sports, Jobs & Taxes: The Economic Impact of Sports Teams and Stadiums,* p. 122.
70. Quoted in John Crompton, "Beyond Economic Impact: An Alternative Rationale for the Public Subsidy of Major League Sports Facilities," *Journal of Sport Management*: 45.
71. George Will, "Modell Sacks Maryland."
72. Veronica Z. Kalich, "A Public Choice Perspective on the Subsidization of Private Industry: A Case Study of Three Cities and Three Stadiums": 214.
73. Quoted in Charles C. Euchner, *Playing the Field: Why Sports Teams Move and Cities Fight to Keep Them,* p. 133.
74. Bob Andelman, *Stadium for Rent: Tampa Bay's Quest for Major League Baseball,* p. xi.
75. Ibid., pp. 91, 93.
76. Quoted in ibid., p. 97.
77. Ibid., p. 99.
78. Richard Corliss, "Build It, and They (Will) MIGHT Come," *Time,* August 24, 1992.
79. Ibid.
80. "Tropicana Field/Tampa Rays," www.ballparkdigest.com
81. Michael Benson, *Ballparks of North America: A Comprehensive Historical Encyclopedia of Baseball Grounds, Yards and Stadiums, 1845 to 1988,* p. 89.
82. Bob Andelman, *Stadium for Rent: Tampa Bay's Quest for Major League Baseball,* p. 7.
83. "White Sox Incentive Deal Striking Out in Legislature," *Chicago Tribune,* May 29, 1988.
84. Charles C. Euchner, *Playing the Field: Why Sports Teams Move and Cities Fight to Keep Them,* p. 134.

85. Ibid., pp. 145–46.
86. Bob Andelman, *Stadium for Rent: Tampa Bay's Quest for Major League Baseball*, p. 7.
87. John P. Pelissero, Beth M. Henschen, and Edward I. Sidlow, "Urban Regimes, Sports Stadiums, and the Politics of Economic Development Agendas in Chicago," *Policy Studies Review*, Vol. 10, Nos. 2/3 (Spring/Summer 1991): 120.
88. Ibid.: 123, 127.
89. Neil deMause and Joanna Cagan, *Field of Schemes: How the Great Stadium Swindle Turns Public Money into Private Profit*, p. 129.
90. Robert A. Baade, "Evaluating Subsidies for Professional Sports in the United States and Europe: A Public-Sector Primer," *Oxford Review of Economic Policy*: 587.
91. John Siegfried and Andew Zimbalist, "The Economics of Sports Facilities and Their Communities," *Journal of Economic Perspectives*, Vol. 14, No. 3 (Summer 2000): 91.
92. Jeff Passan, "Marlin's profit came at taxpayer expense," www.sports.yahoo.com, August 24, 2010.
93. George Lipsitz, "Sports Stadia and Urban Development: A Tale of Three Cities," *Journal of Sport and Social Issues*: 1.
94. Ibid.: 3–4.
95. Lawrence O. Christensen, "August Busch Jr.," *Dictionary of Missouri Biography* (Columbia: University of Missouri, 1999), p. 137.
96. Michael Benson, *Ballparks of North America: A Comprehensive Historical Encyclopedia of Baseball Grounds, Yards and Stadiums, 1845 to 1988*, p. 348.
97. George Lipsitz, "Sports Stadia and Urban Development: A Tale of Three Cities," *Journal of Sport and Social Issues*: 4.
98. Richard Temple Middleton IV, "The Politics of Stadium Development in Phoenix, Arizona," in *The Economics and Politics of Sports Facilities*, pp. 107–108.
99. Dave Zirin, "Football in L.A.," *Los Angeles Times*, October 29, 2009.
100. Anthony York, "Push for NFL Stadium in Los Angeles resurfaces at session's 11th hour," *Capitol Weekly*, September 3, 2009.
101. Jacob Adelman, "LA stadium builder: Jaguars, Bills first on list," Associated Press, January 5, 2010.
102. Gary Cartwright, "A Barnum Named Hofheinz, a Big Top Called Astrodome," *New York Times Magazine*, July 21, 1968.
103. George Lipsitz, "Sports Stadia and Urban Development: A Tale of Three Cities," *Journal of Sport and Social Issues*: 10.
104. Katherine C. Leone, "No Team, No Peace: Franchise Free Agency in the National Football League," *Columbia Law Review*: n317.
105. Gary Cartwright, "A Barnum Named Hofheinz, a Big Top Called Astrodome," *New York Times Magazine*.
106. Ibid.
107. George Lipsitz, "Sports Stadia and Urban Development: A Tale of Three Cities," *Journal of Sport and Social Issues*: 12.
108. Gary Cartwright, "A Barnum Named Hofheinz, a Big Top Called Astrodome," *New York Times Magazine*.
109. Richard Williamson, "All In for Houston Agency?" *The Bond Buyer*, December 22, 2009, www.bondbuyer.com.

110. George Will, "Modell Sacks Maryland," *Newsweek.*
111. Marc Levinson, "Fields of Schemes," *Newsweek,* December 11, 1995.
112. Raymond J. Keating, "The NFL Oilers: A Case Study in Corporate Welfare," *The Freeman,* Vol. 48, Issue 4, April 1998.
113. Ibid.
114. Ibid.
115. Katherine C. Leone, "No Team, No Peace: Franchise Free Agency in the National Football League," *Columbia Law Review*: n84.
116. Richard Sandomir, "A Texas-Size Stadium," *New York Times,* July 17, 2009. See also Bruce K. Johnson, Michael J. Mondello, and John C. Whitehead, "Contingent Valuation of Sports: Temporal Embedding and Ordering Effects," *Journal of Sports Economics,* Vol. 7, No. 3 (2006): 267–88.
117. Robert Wilonsky, "Going Deep: The Cowboys want to move back to Fair Park. so there goes the neighborhood?" Dallas *Observer,* May 27, 2004.
118. Quoted in John Crompton, "Beyond Economic Impact: An Alternative Rationale for the Public Subsidy of Major League Sports Facilities," *Journal of Sport Management*: 44.
119. David Wethe and Sally Claunch, "Arlington, Texas, council begins weighing Cowboys stadium deal," Fort Worth *Star-Telegram,* July 28, 2004.
120. David Wethe, "Dallas Cowboys and Arlington, Texas, council near stadium deal," Fort Worth *Star-Telegram,* August 11, 2004.
121. Jeff Mosier, "Arlington to help fund Cowboys stadium: Voters approve tax increases for $650 million venue," Dallas *Morning News,* November 3, 2004.
122. Mitchell Schnurman, "Looking Back at Stadium Victory," Fort Worth *Star-Telegram,* November 7, 2004.
123. Jeff Mosier, "Moving out of stadium's way: Family last to exit Arlington neighborhood set to be razed," Dallas *Morning News,* February 7, 2006.
124. Jeff Mosier, "Settlement ends stadium land acquisition," Dallas *Morning News,* March 25, 2009.
125. Linda Lancaster, "Lemmings and Toll Roads," Dallas *Observer,* April 5, 2007.
126. Richard Sandomir, "A Texas-Size Stadium," *New York Times.*
127. "Tex. s Eminent Domain Reform Weakened Minutes Before Passage," States News Service, June 1, 2009.
128. Richard Sandomir, "A Texas-Size Stadium," *New York Times.*
129. Neil deMause and Joanna Cagan, *Field of Schemes: How the Great Stadium Swindle Turns Public Money into Private Profit,* p. 66.
130. Dean V. Baim, *The Sports Stadium as a Municipal Investment,* p. 135.
131. Ibid., p. 139.
132. Robert K. Whelan and Alma H. Young, "The Politics of Planning and Developing New Sports Facilities: The Case of Zephyrs Park and the New Orleans Arena," in *The Economics and Politics of Sports Facilities,* p. 112.
133. James Quirk and Rodney D. Fort, *Pay Dirt: The Business of Professional Sports Teams* (Princeton: Princeton University Press, 1997), p. 158.
134. Dean V. Baim *The Sports Stadium as a Municipal Investment,* p. 122.
135. Quoted in Clyde Brown and David M. Paul, "Local Organized Interests and the 1996 Cincinnati Sports Stadia Tax Referendum," *Journal of Sport and Social Issues,* Vol. 23, No. 2 (May 1999): 218.

136. By one estimate, bond lawyers who act as consultants in the issuance of bonds generally have fees of over $1 million. John L. Crompton, Dennis R. Howard, and Turgut Var, "Financing Major League Facilities: Status, Evolution and Conflicting Forces," *Journal of Sport Management*, Vol. 17, No. 2 (April 2003): 178.
137. Clyde Brown and David M. Paul, "Local Organized Interests and the 1996 Cincinnati Sports Stadia Tax Referendum," *Journal of Sport and Social Issues*: 222.
138. Ibid.: 223.
139. Ibid.: 225–26.
140. John Helyar, "Ball Control: A City's Self-Image Confronts Tax Revolts in Battle on Stadiums," *Wall Street Journal*, March 19, 1996.
141. Clyde Brown and David M. Paul, "Local Organized Interests and the 1996 Cincinnati Sports Stadia Tax Referendum," *Journal of Sport and Social Issues*: 227–28.
142. Ibid.: 231–32.
143. John Helyar, "Ball Control: A City's Self-Image Confronts Tax Revolts in Battle on Stadiums," *Wall Street Journal*.
144. Clyde Brown and David M. Paul, "Local Organized Interests and the 1996 Cincinnati Sports Stadia Tax Referendum," *Journal of Sport and Social Issues*: 230.
145. Ibid.: 229.
146. John Helyar, "Ball Control: A City's Self-Image Confronts Tax Revolts in Battle on Stadiums," *Wall Street Journal*.
147. Clyde Brown and David M. Paul, "Local Organized Interests and the 1996 Cincinnati Sports Stadia Tax Referendum," *Journal of Sport and Social Issues*: 229.
148. Ken Belson, "Stadium Boom Deepens Municipal Woes," *New York Times*, December 24, 2009.
149. Glen Gendzel, "Competitive Boosterism: How Milwaukee Lost the Braves," *Business History Review*: 564.
150. Joseph L. Bast, "Sports Stadium Madness: Why It Started/How to Stop It," Heartland Institute of Chicago, Policy Study No. 85, February 23, 1998, p. 16.
151. George H. Sage, "Stealing Home: Political, Economic, and Media Power and a Publicly-Funded Baseball Stadium in Denver," *Journal of Sport and Social Issues*, Vol. 17, No. 2 (August 1993): 110.
152. Ibid.: 115.
153. William Fulton, "Politicians Who Chase After Sports Franchises May Get Less Than They Pay For," *Governing*, Vol. 1, No. 6 (March 1988): 38.
154. George H. Sage, "Stealing Home: Political, Economic, and Media Power and a Publicly-Funded Baseball Stadium in Denver," *Journal of Sport and Social Issues*: 115–16.
155. Ibid.: 116.
156. Ibid.: 118.
157. John C. Melaniphy, "The Impact of Stadiums and Arenas," *Real Estate Issues*, Vol. 21, No. 3 (December 1996): 36.
158. Peter Navarro, "San Diego latest pawn in stadium blackmail," *USA Today*, February 20, 1997.
159. "Tailgating May Disappear at Chargers' New Stadium," MediaVentures, February 11, 2010, www.football.ballparks.com.

160. "Chargers Considering Roof for Stadium," MediaVentures, April 22, 2010, www.football.ballparks.com.
161. Erle Norton, "Fourth and Goal. Football at Any Cost," *Wall Street Journal,* October 13, 1993.
162. Ibid.
163. John Crompton, "Beyond Economic Impact: An Alternative Rationale for the Public Subsidy of Major League Sports Facilities": 43.
164. Erle Norton, "Fourth and Goal. Football at Any Cost," *Wall Street Journal.*
165. Robert A. Baade and Allen R. Sanderson, "The Employment Effect of Teams and Sports Facilities," in *Sports, Jobs & Taxes: The Economic Impact of Sports Teams and Stadiums,* p. 94.
166. Ken Belson, "A New N.F.L. Stadium, But at Whose Cost?" *New York Times,* October 11, 2009.
167. Ken Belson, "As Stadiums Vanish, Their Debt Lives On," *New York Times,* September 7, 2010.
168. Gregg Easterbrook, "Two Games are Enough; Tear it Down," Tuesday Morning Quarterback, September 28, 2010, www.sports.espn.go.com.
169. Jay Weiner, *Stadium Games: Fifty Years of Big League Greed and Bush League Boondoggles* (Minneapolis: University of Minnesota Press, 2000), p. xvii.
170. Ibid., pp. 251–53.
171. Neil deMause and Joanna Cagan, *Field of Schemes: How the Great Stadium Swindle Turns Public Money into Private Profit,* p. 36.
172. Ibid., p. 208.
173. "Vikings Would Pay for Third of Stadium," Associated Press, January 13, 2011.
174. Dave Zirin, "Silver Lining for Vikings Fans (Politically)," January 26, 2010, www.TheNation.com.
175. Robert Cribb, "Ziegler's NHL dream got burned in Sunbelt," Toronto *Star,* October 3, 2009.
176. James Mirtle, "NHL attendance dips by 2.5 per cent," Toronto *Globe and Mail,* June 15, 2010.
177. Tripp Mickle, "Sun Belt hot and cold for NHL," *Sports Business Journal,* Oct 5, 2009, www.sportsbusinessjournal.com.
178. James Quirk and Rodney D. Fort, *Pay Dirt: The Business of Professional Sports Teams,* p. 30.
179. Tim Wendel, *Buffalo, Home of the Braves* (Traverse City, MI: SunBear Press, 2009), p. 211.
180. Steven A. Riess, "Historical Perspectives on Sports and Public Policy," in *The Economics and Politics of Sports Facilities,* p. 32.
181. Dennis Coates and Brad R. Humphreys, "Do Economists Reach a Conclusion on Subsidies for Sports Franchises, Stadiums, and Mega-Events?" *Econ Journal Watch,* Vol. 5, No. 3 (September 2008): 297–98.
182. George Will, "Avaricious developers and governments twist the meaning of 'blight,'" *Washington Post,* January 3, 2010.
183. Damon W. Root, "When Public Power is Used for Private Gain," www.reason.com, October 8, 2009.

184. Alex Williams, "Back to the Future," *New York*, October 27, 2003.
185. Ibid.
186. George Will, "Avaricious developers and governments twist the meaning of 'blight,'" *Washington Post*.
187. Nicholas Confessore, "Perspectives on the Atlantic Yards Development Through the Prism of Race," *New York Times*, November 12, 2006.
188. Alex Williams, "Back to the Future," *New York*.
189. Neil deMause and Joanna Cagan, *Field of Schemes: How the Great Stadium Swindle Turns Public Money into Private Profit*, p. 288.
190. Nicholas Confessore, "Perspectives on the Atlantic Yards Development Through the Prism of Race," *New York Times*.
191. www.atlanticyardsreport.blogspot.com, November 3, 2010.
192. Sewell Chan, "Report Sees Loss in Brooklyn Arena," *New York Times*, September 11, 2009.
193. Charles V. Bagli, "$500 Million in Bonds for Nets' Arena Sell in 2 Hours," *New York Times*, December 16, 2009.
194. Andy Newman and Charles V. Bagli, "Daniel Goldstein, Last Atlantic Yards Holdout, Leaves for $3 Million," *New York Times*, April 21, 2010.

Chapter 5

If You Build It, Prosperity Will Not Come: What the Studies Say

The Rich Get Richer…

Okay, so we've heard the case histories. What about academic studies of the question? Is there a consensus?

Oh, is there ever a consensus…

"With the exception of the racial ban that kept blacks from the game until 1947, the publicly financed stadium is the most regrettable feature of the business of baseball," says Neil J. Sullivan, author of *The Dodgers Move West*.[1] A strong assertion — but one that applies, perhaps with even greater force, to those sports that lacked a statutory color barrier: football, basketball, and hockey.

There are, as of 2012, 122 teams in the National Football League (NFL), Major League Baseball (MLB), the National Basketball Association (NBA), and the National Hockey League (NHL). They play in 60 stadiums (the Oakland and Miami football and baseball teams share their parks) and 50 arenas (hockey and basketball teams share venues in Los Angeles, New York, Philadelphia, Dallas, Washington, Atlanta, Denver, Boston, and Chicago). Neil deMause, in his introduction to the updated edition of *Field of Schemes: How the Great Stadium Swindle Turns Public Money into Private Profit* (2008), estimates that the annual subsidy to stadium/arena construction is now two billion dollars.[2] Although flat-out 100% municipal construction and ownership is now rarer than in the "pave it and they will come" 1950s and 1960s, the magnitude of the subsidies shows no sign of contraction.

J.T. Bennett, *They Play, You Pay: Why Taxpayers Build Ballparks, Stadiums, and Arenas for Billionaire Owners and Millionaire Players*, DOI 10.1007/978-1-4614-3332-3_5,

What really galls many taxpayers is the reverse Robin Hood quality of such subsidies. After all, these are not grants to Little Leagues or Boys Club boxing tournaments or girls' softball fields. No, the men who reap the largest rewards from sports-related public policy in this country are, in many cases, the richest men in the country. As Mark S. Rosentraub writes in *Major League Losers: The Real Cost of Sports and Who's Paying for It*, "A welfare system exists in this country that transfers hundreds of millions of dollars from taxpayers to wealthy investors and their extraordinarily well-paid employees" — even though sports have a "minuscule" impact on a city's economy.[3]

In 2010, the average salary in the major sports leagues was: Baseball: $3.015 million[4]; Football: $1.896 million[5]; Basketball: $3.4 million[6]; and Hockey: $2.4 million.[7] The median household income in America, by puny contrast, is $50,221.[8] As Robert A. Baade has written, "Taxing low or average income groups to enhance the financial privilege of unusually wealthy professional athletes raises ethical and moral concerns."[9] And stadium subsidies boost athlete's salaries to levels higher than they would be in the absence of subsidies.

As for the owners...Croesus might have to strain to join this group.

Lest anyone think that the flamboyant Dallas Cowboys' owner Jerry Jones is an exception, and that the other owners in the NFL are far from the super-wealthy, and instead are for the most part modestly well-off sportsmen who in purchasing and maintaining a professional football team are performing a public benefaction to their hometowns — "giving back," as they like to say — *Forbes*, the self-described "capitalist tool," has the facts.

As of the 2010 football season, more than half of the owners in the NFL (16 of 31, with Green Bay's community-owned team the 32nd team) were *billionaires*. Not millionaires — that status is reserved for mere players. But billionaires. You would think that accumulating such wealth would more or less disqualify one from public alms — but you would be wrong.

Topping the *Forbes* list of the billionaire football owners is Paul Allen, cofounder of Microsoft and owner of the perennially mediocre Seattle Seahawks. His net worth was pegged at $12.7 billion, give or take a few shekels.[10]

The other 15 billionaire owners, with the sources of their wealth, their teams, and their net worths, are:

- Stephen Ross, real estate development, Miami Dolphins, $3.1 billion
- Stan Kroenke, real estate development (and marrying a Walton — of Benton, Arkansas, not Walton's Mountain), St. Louis Rams, $2.7 billion
- Malcolm Glazer, food service, Tampa Bay Buccaneers, $2.6 billion

- Jerry Jones, oil and gas exploration, Dallas Cowboys, $2 billion
- Robert Kraft, paper products, New England Patriots, $1.5 billion
- James Irsay, inheriting a football team, Indianapolis Colts, $1.4 billion
- Robert McNair, energy, Houston Texans, $1.4 billion
- Stephen Bisciotti, staffing company, Baltimore Ravens, $1.3 billion
- Arthur Blank, Home Depot, Atlanta Falcons, $1.2 billion
- Bud Adams, oil, Tennessee Titans, $1.15 billion. (As *Forbes* points out, Adams bought the team, then the Houston Oilers, for $25,000 in 1960, and even "after factoring in 50 years of inflation, Adams made over 5,400 times his money." Not a bad investment.)
- Alex Spanos, real estate development, San Diego Chargers, $1.1 billion
- Dan Snyder, communications, Washington Redskins, $1.1 billion. (Snyder's ineptness as an owner has made the Redskins a laughingstock, though at least his team's home, FedExField in Landover, Maryland, was built primarily with private funds. Note that Snyder's predecessor as owner, Jack Kent Cooke, for whom the stadium was briefly named, didn't stick his hand in the cookie jar. In 1992, Cooke and Virginia Governor Douglas Wilder (D) negotiated a plan under which the state of Virginia would finance $130 million of road, rail line, and infrastructure improvements leading into a Redskins stadium in Alexandria, Virginia. The deal seemed to be set — until outraged citizens in Alexandria, peeved that this giveaway to the unlikable Cooke would rip up large parts of their city, registered their disapproval loudly enough to derail the plan. And although FedExField is sometimes called a privately built stadium, Robert Baade notes that Maryland "contributed $78 million to the project for access roads, parking lots, and other facilities around the stadium."[11])
- Tom Benson, automobile dealership, New Orleans Saints, $1.05 billion
- Jeff Lurie, film producer, Philadelphia Eagles, $1 billion
- William Clay Ford Sr., being Henry Ford's grandson, Detroit Lions, $1 billion[12]

What hard-hearted wretch could object to lavishing public monies upon such men?

The symbolism of such subsidies calls forth a howls of populist protest. Government giveaways to the rich! Spoiled athletes and billionaire owners are feeding at the public trough, while hard-pressed taxpayers have to scrimp and save to pay their property taxes and health insurance. As Allen R. Sanderson of the University of Chicago's Department of Economics writes, the "primary beneficiaries of stadium reconstruction and team relocations" are the "cartel leagues, team owners, players, and diehard fans, while the losers are taxpayers."[13]

In "Subsidizing Stadiums," tax-law specialist Dennis Zimmerman writes that "the incidence of private consumption benefits is becoming more and more concentrated among the middle- and upper-income segments of the local population and business community."[14] The working stiff pulling a double shift at 7–11 is paying taxes that build a stadium whose luxury suites are occupied by CEOs who make as much in a fortnight as the Slurpee-pourer makes in a year. Lives there a political philosopher who will defend this arrangement?

Billions and Billions

The size of the subsidies has been measured most recently by Mayya M. Komisarchik and Aju J. Fenn of the Colorado College Department of Economics and Business. In an April 2010 working paper, Komisarchik and Fenn relay their findings that between 1995 and 2009, governments contributed $10.34 billion to major league stadium and arena projects, while private sources were responsible for $7.08 billion. The respective percentages are 59 and 41. Broken down by sport, the public share of new football stadiums was 59% ($4.03 billion public, $2.82 billion private); the public share of new baseball parks was 59% ($3.15 billion public, $2.19 billion private); and the public share of new basketball and hockey arenas — some shared venues, some not — was 60% ($3.11 billion public, $2.07 billion private). New stadiums and arenas are going up at a faster rate than ever before, such that the 20 years span between 1995 and 2015 will see "the highest sums of public dollars ever spent on major league sports."[15]

While innovations in stadium design have in recent years enabled teams to virtually bathe in a deepening pool of revenues from such sources as private seat licenses, luxury suite rentals, stadium naming rights, parking, and concessions so diverse that word *concessions* — which calls up images of shriveled hot dogs and warm beer — hardly do them justice — owners have not exactly demanded that they be removed from the government teat. (More likely they demand a greater share of revenues from such sources — even when the park and parking lot have been built with government funds.) As Komisarchik and Fenn note, "94% of all MLB ballparks are at least partially funded by public dollars, as are 95% of NFL stadiums, 86% of dual use NBA/NHL arenas, 94% of NBA arenas and 88% of arenas serving the NHL."[16]

Stadiums and arenas are also being built at an unprecedented rate: whereas before 1990, "Major league athletic venues were built at a rate of 0.36 arenas per year, or one arena every 2.8 years," the rate since 1990 is "comparably astounding at 3.63 stadiums per year."[17] In fact, over half of the teams in the NFL, NHL, NBA, and MLB have "moved into a new or substantially renovated venue since 1990."[18]

Strange as it may seem in this era of "public–private partnerships" under which states and municipalities build ballparks and stadiums for *Forbes* 400 owners, in pre-New Deal America, as we have seen, such edifices of sport were almost always constructed on privately assembled land at the expense of team owners or entrepreneurs. As late as 1950, Cleveland was the only MLB team to play in a government-owned park. Baseball's National League had no government-built ballparks at the time. Nor did any of the six NHL teams skate on state-supplied ice, though a substantial minority of teams in the NFL — 36% — and NBA — 46% — played in publicly owned stadiums or arenas in 1950.[19]

As economists James Quirk of Cal Tech and Rodney D. Fort of Washington State write in *Pay Dirt: The Business of Professional Sports Teams*: "It seems a long time ago now, but it was only forty years ago that local governments simply were not players in the pro sports team game. Partly, this was because of the ethos of the times; local governments weren't supposed to be in the business of subsidizing private enterprises. But added to this was the limited political clout of the sport constituency, which typically did not include the local establishment."[20]

"In the era just after World War II," they continue, "most major league teams in all sports played in privately owned stadiums or arenas. Most NFL teams played in baseball parks that were owned by baseball teams; and among MLB teams, only Cleveland played in a publicly owned stadium. NHL teams played in privately owned arenas, and about half of the NBA teams also played in privately owned arenas."[21] And remember, the late 1940s were hardly some laissez-faire Eden in American economic history. The Second World War had just ended, and it had achieved an unprecedented concentration of economic and political power in Washington, DC. Unlike after the First World War, the demobilization was modest and limited; a Cold War with the Soviet Union would keep the government bureaus humming, and most aspects and programs of the New Deal remained long after the Great Depression had ended.

Still, few would have contended in the late 1940s that building stadiums or ballparks or arenas for football or baseball or basketball or hockey teams was a legitimate function of city, state, or federal government.

The apex of the public subsidy era occurred roughly from 1961 through 1984, which corresponded, in a noted noncoincidence, with the advent of the Great Society and the expansion of welfare entitlements not only for the "poor" but also for the middle class. The vast majority of parks and stadiums in this era were built almost entirely by taxpayer subsidy. That these venues were often characterless, boring, sterile multipurpose facilities, as we have seen, was perhaps to be expected. Modern government architecture has seldom satisfied the soul.

What Texas A&M professor John Crompton terms the "transitional" era of stadium funding ran from 1985 to 1994 and was marked by a mix of private and public funding (though public still dominated, as it does today) for "elaborate, 'fully loaded,' single sport facilities." The ballparks were better — think Baltimore's Camden Yards versus Cincinnati's Riverfront Stadium — and the private share of construction was, percentage-wise at least, greater. (Camden Yards, ironically, was an exception.) Basketball and hockey arenas, too, were taking shape as public/private partnerships, as for instance with Phoenix's America West Arena.

Since 1995, in what Crompton dubs the "Fully-Loaded (Private–Public Partnership)" era, the cost of facilities has escalated greatly, and while the franchises themselves are often contributing a substantial sum toward construction, overall costs to the taxpayer have continued to rise. Now-common features such as luxury suites and retractable roofs add considerably to construction costs. So while "the proportionate investment of public resources into facilities may have declined, the net annual public subsidy to the franchises often has increased."[22] Privatization may have achieved a cachet among policymakers in the late 1980s, but its virtues were not widely trumpeted in sports policy. By 1991, "65 of the 84 stadiums and arenas in use (77 percent) were publicly owned."[23]

Still, especially with arenas, in recent years "there has been a dramatic shift toward franchise owners assuming a greater share of development costs of the venues in which their teams play," write Crompton and Turgut Var of Texas A&M and Dennis R. Howard of the University of Oregon. The period of the early 1960s through the early 1980s was marked by a degree of 100% government funding that would have been the envy of Bulgarian economic planners. In the Great Society, even millionaire sports moguls were eligible for a free lunch. That began to change in the mid-1980s — whether through the influence of ascendant free-market doctrines or just because municipal finances were running low — and yet while a private component is now often — though often not — expected in any new construction, the public contribution to these increasingly expensive entertainment venues is hardly shrinking. In fact, note Crompton, Var, and Howard, "in real dollar terms" the government share of the cost of arena construction is actually higher today, in the supposedly golden age of private–public partnership, than it was in the 1970s, the age of the free lunch.[24]

The advent of the fully loaded stadium rocketed the cost of construction substantially. Andrew Moylan, in a study for the National Taxpayers Union Foundation, found that such costs were 60% higher a decade after the fully loaded era got cranking around 1990. Noting that the cost of materials such as iron and steel had remained relatively stable in that period, Moylan opined that the probable reason for the sharp increase in the price of a new stadium was the proliferation of luxury suites, in-stadium restaurants and bars, close-to-the-action "club-level" seating, and other features that by 2000 had become de rigueur in the spectating life.[25]

Moylan also found, in his study of 53 stadiums and arenas built for NFL, MLB, and NBA teams between 1990 and 2004, that "stadiums that were constructed with 50% or more in taxpayer subsidies were $65 million more expensive on average than those constructed with less than 50% in subsidies." Why? Well, cost-containment incentives tend to be stricter when one payer shoulders the cost himself rather than passing it along to a third party. Interestingly, the most expensive stadiums were not those with 100% public financing but, rather, those whose public share of the bill came to between 60% and 79.9%. Moylan theorizes that government officials seek to "minimize public outrage" over cost overruns and astronomical corporate welfare packages, but the soothing sound of a "public–private partnership" will lull watchdogs to sleep.[26]

Moreover, the numbers thrown around in stadium debates may well understate the true cost to taxpayers. Judith Grant Long, a professor of Urban Planning and Policy Development at Rutgers, writes in the *Journal of Sports Economics* that "most existing estimates of public subsidies for sports facilities are significantly underestimated."

Undertaking the admittedly "cumbersome data-collection process" entailed in assessing the actual scope of such subsidies, Professor Long found that land and infrastructure improvement costs are typically omitted from these calculations, as are the costs of municipal services, capital improvements, and foregone property taxes. Examining the 99 major league sports facilities in use over the course of 2001, Long pegged the unreported public subsidies at $5 billion, or an average of $50 million per facility. Broken down by sport, the average public subsidy of a MLB park was $53 million. Thus, the actual public subsidy of ballparks in 2001 was not the $165 million reported by industry sources, but rather $218 million. The underreported public subsidies for the other three major sports were $53 million for an NBA arena, $46 million for an NHL arena, and $41 million for an NFL stadium.[27]

Adding in these stealth subsidies makes even clearer just how government-dependent the major league sports industry is. While industry sources assert that 56% of the cost of the average new sports facility is government-subsidized, Long notes that "my findings show that after adjusting for omitted subsidies, the average public share is 79% — an increase of 23 percentage points."[28]

The pace of construction has reached a sprint today, certainly in comparison with the leisurely stroll of the 1970s, when several of the most charmless multipurpose cookie-cutters were built, almost always with a heavy or even exclusive public financing component. And yet, when polled, the American people profess to oppose giveaways to the Jerry Joneses and Dan Snyders of the sports world. Public opinion surveys have shown that up to 80% of those responding are against the use of taxpayer money to construct stadiums or ballparks for major league teams.[29]

Why do cities subsidize stadiums? Largely due to the "widespread belief that sports facilities are an engine of local economic development" — a view wildly out of tune with reality, as virtually every economist who has ever studied the matter has concluded.[30]

So why not just give a team a cash subsidy rather than build a stadium? As John Siegfried and Andrew Zimbalist answered this hypothetical in the *Journal of Economic Perspectives*, the reasons are legion. Building a stadium gives the politicos carrots to distribute to unions, contractors, and other influential parties. It enables a city to tie a team thereto through a long-term (if often loophole-ridden) lease. By putting the subsidy into material form as a stadium, it discourages other subsidy-seekers from pressing their claims. And it deflects criticism over corporate welfare, for "direct cash subsidies delivered from government to wealthy team owners in full view of voters are unhealthy for local politicians."[31] Flat-out cash giveaways send out a very bad vibe to voters.

Let us rephrase the question. Why do politicians succumb so easily to the meretricious lures of the subsidy seekers? For one thing, they are politicians, whose job is to spend other people's money. If they can bask in reflected glory, all the better. When Coleman Young, Detroit mayor, began speaking of a downtown domed stadium in the 1980s, a city councilman said, "I think what Coleman Young is looking for is a stadium named Coleman A. Young Stadium."[32] Politicians want credit for constructing stadiums and arenas, attracting or keeping professional sports teams, but they don't want the blame for tax increases. So the best of all possible worlds is for the voters to tax themselves via referendum — or the owner to build it himself.

But beyond that, stadiums are the kind of big, sexy, gaudy public-works projects that grab headlines and dominate skylines and enhance a city's reputation as "World Class" or "First Class" or "Major League" or whichever boosterish adjective is being currently applied.

"Do we really want to be like Louisville?" asked a Cincinnati politician trying to explain why he supported taxpayer subsidies to build new stadiums for the Reds and the Bengals.[33] The price of not descending to such horrific status was $540 million. How much better, though, is Cincinnati's national image, associated as it is with the perennially bad and often thuggish Bengals, than that of Louisville, most famous for hosting the Kentucky Derby at Churchill Downs, which was not built by levies on the taxpayers of the city and state but rather by the sale of 320 membership subscriptions at $100 each in 1875 by Col. M. Lewis Clark?

Enough Empirical Evidence to Fill Yankee Stadium?

Economists are not noted for singing harmony on many issues, but on the public subsidies for sports facilities question, at least, they are in near-unanimity. When economists Dennis Coates of the University of Maryland and Brad R. Humphreys of the University of Alberta in Canada undertook to answer for *Econ Journal Watch* the question, "Do Economists Reach a Conclusion on Subsidies for Sports Franchises, Stadiums, and Mega-Events?" the overwhelming evidence pointed to *Yes*. In fact, as they note, in 2005 Robert Whaples had surveyed a random sample of members of the American Economic Association, asking for agreement or disagreement with the statement "Local and state governments in the U.S. should eliminate subsidies to professional sports franchises."

Fifty-eight percent answered "strongly agree," 28% agreed, 10% were neutral, and a paltry 5% disagreed — perhaps just to be disagreeable, for this "exceptional consensus" featured the highest "strongly agree" percentage of any of the twenty policy questions on which Whaples polled the membership.[34]

"The clear consensus among academic economists," write Coates and Humphreys, "is that professional sports franchises and facilities generate no 'tangible' economic impacts in terms of income or job creation and are not, therefore, powerful instruments for fostering local economic development." We have over 20 years of such research on this subject, and the findings are "strikingly consistent" in asserting "almost no evidence that professional sports franchises and facilities have a measurable impact on the economy."[35]

As Mark Rosentraub found when measuring the contribution of sports to the economy of Arlington, Texas — hardly a blue-collar smokestack town — "manufacturing is substantially more important than the entertainment, sports-related, and restaurant businesses combined." He said, "Sports teams are small to medium-sized businesses, and even in those areas with several teams, the professional team sports component of an economy never accounts for as much as 1% of the jobs or 1% of the payrolls in that county."[36]

Or as economist Roger Noll of Stanford says, "Opening a branch of Macy's has a greater economic impact" than opening a new ballpark or stadium.[37] And Macy's doesn't ask the host city to build a store for them.

Also unlike Macy's, a football stadium is used 10 days a year — 12 if the team is exceptionally good. A baseball stadium is used for 80 days — 90 if the team is of World Series caliber. And in the case of arenas, the justification for public subsidy is even weaker. "With arenas," Andrew Zimbalist says, "as opposed to outdoor football stadiums, you ought to be able to get much more private financing. The basketball team alone gives you 41 dates, and that's even if [an NHL team doesn't] come. A football stadium gives you ten dates, and there's not much else you can do there to create revenue except maybe have Billy Graham come and speak, or Bruce Springsteen give a concert." Madison Square Garden, the privately owned multipurpose arena in Manhattan, hosts upwards of 275 events per year.[38]

You could probably fill a good-sized display shelf at Macy's with the research on the benefits, or lack thereof, of publicly funded sports venues. Consider these studies (proponents of stadium subsidies sure don't).

In 2004, economists Coates and Humphreys warned that "Our own research suggests that professional sports may actually be a drain on local economies rather than an engine of economic growth."

Their research encompassed each of the 37 cities that hosted a MLB, football, or basketball team between 1969 and 1996. They found, among other things, that:

- "The presence of pro sports teams" in those 37 cities "had no measurable positive impact on the overall growth rate of real per capita income in those areas." In fact, it had a "statistically significant *negative* impact" on per capita income. [My emphasis]

- The presence of major league teams "had a statistically significant negative impact on the retail and service sectors of the local economy." The net loss of jobs in the service sector averaged 1,924. While the presence of these teams did boost "wages in the hotels and other lodgings sector by about $10 per year," it decreased "wages per worker in eating and drinking establishments by about $162 per year."
- The employment sector that showed the largest average increase due to the presence of major league teams was the amusements and recreation field, whose increase of $490 per year is due to the often astronomical salaries of the athletes themselves, which have something of a distortive effect. Remove the handful of athletes and the increase virtually disappears.

Looking specifically at baseball, Coates and Humphreys found that bunters weren't the only ones to sacrifice when America's Game came to town. For "on average, professional baseball lowered the earnings of workers in eating and drinking establishments by about $144 per employee per year. Baseball also lowered the per employee annual earnings of workers in the hotel and lodging sector by about $38" and, starkly, it reduced average annual earnings in the amusements and recreation field — and recall, this includes the huge salaries of the players themselves — by $503.[39]

Clearly, note the authors, there is a substitution effect at work here. Dollars spent on baseball are dollars *not* spent in local bars or nightclubs or in bowling alleys or even, perhaps, at the opera. This is, of course, consistent with decades of research in sports economics, and yet as familiar as these findings are to economists, they never seem to penetrate either the local newspapers, which almost invariably flack for new stadiums, or the Chamber of Commerce booster set, which is gung-ho to proclaim their particular city "major league."

As Mark S. Rosentraub and David Swindell note, a baseball team can increase economic activity in four ways: (1) by attracting "recreational spending away from other geographical areas"; (2) by "deflect[ing] residents from going elsewhere for baseball games or recreation"; (3) by bringing in visiting teams, which use a city's hotels and bars and restaurants; and (4) if its presence "actually increases aggregate spending," as consumers decrease savings or increase their earnings in order to buy tickets and patronize the ballpark.[40]

Other sources of economic growth, as Rosentraub and Swindell point out, can be the construction necessary to build or improve the park, the creation of permanent and seasonal jobs at the ballpark, and increased tax revenues.

There is also the hard to quantify but undeniable psychic benefit of having a team, which can enhance a community's reputation and draw its members together. Civic pride is important to a place and its residents, but is it really necessary to build a multibillion dollar stadium just to foster it?

(Steven A. Riess, after studying the effect that baseball stadiums in New York City and Chicago had on the surrounding area in the late nineteenth and early twentieth centuries, concluded that "the fields did not have much of an impact on an area's future other than a psychological one since their influence on land uses and property values seldom extended more than a couple of blocks from their entrance."[41] So even in the days of the sub-$1 million stadium paid for entirely out of the pocket of an owner and his associates, the economic benefits of a MLB park were strictly limited.)

The modern jobs tally is no more impressive. "The experience of a cross section of cities in the United States during the past quarter century shows scant evidence that professional sports create a significant number of new jobs," write Robert A. Baade and Allen R. Sanderson.[42]

In an earlier study, Baade looked at the 36 metropolitan areas that had a team in one of the four major professional sports — football, baseball, basketball, or hockey — between 1958 and 1987, as well as 12 metropolitan areas that had no teams in that period.

He found that in 30 of the 32 cities in which there was a change in the number of professional sports teams, there was "no significant relationship between the presence of the teams and real, trend-adjusted, per capita personal income growth." The two exceptions were Indianapolis, which had a significantly positive relationship — though you will recall the limits of the Indianapolis example from the previous chapter — and Baltimore, where the relationship was significantly negative.

Nor was there a significant relationship between real, trend-adjusted, per capita income growth, and the presence of a stadium in 27 of the 30 cities in which the number of stadiums or arenas newer than 10 years old showed a change. In the other three metropolitan areas (St. Louis, San Francisco/Oakland, and Washington, DC), the relationship was negative. Baade speculates that in the cases of the Bay Area and the nation's capital, the contrast was between the well-paid workforce of these affluent areas and the often low-paying stadium jobs. Ushers, parking lot attendants, and concession-stand cashiers do not make nearly the wages earned by, say, a San Francisco architect or a Beltway lobbyist.[43]

In any case, Baade concludes that "public funding of professional sports stadiums is not a sound civic economic investment." In fact, he muses that "a sound development strategy might simply leave these 'development' funds in the private economy. Private spending has spin-off benefits of its own, and there is reason to believe that private individuals and businesses often make better spending decisions in the private sector than public officials do in the public sector."[44]

Baade and Richard F. Dye explored "The Impact of Stadiums and Professional Sports on Metropolitan Area Development" in a paper in the journal *Growth and Change*. Applying regression analysis to census data from nine metropolitan areas with MLB or football teams — Cincinnati, Denver, Detroit, Kansas City, New Orleans, Pittsburgh, San Diego, Seattle, and Tampa Bay — Baade and Dye concluded that "the presence of a new or renovated stadium has an uncertain impact on the levels of economic activity and possibly a negative impact on local development relative to the region."[45] Stadium subsidies, they say, "might bias local development toward low-wage jobs."[46]

In fact, in four of the metropolitan areas under study (Cincinnati, Detroit, Kansas City, and Tampa Bay), "stadium construction or renovation is significantly correlated with a reduction" of the area's share of regional income. There was a significantly positive correlation in two areas, New Orleans and Seattle. Besides the very small number of high-salaried jobs a football or baseball team brings to a region — and those high-salaried jobs are almost invariably held by athletes from outside the area — the presence of such teams and the stadiums in which they perform tends to "divert economic development toward labor-intensive, relatively unskilled labor (low-wage) activities."[47] (According to Vanderbilt University economist John Siegfried, over 70% of NBA players live outside the metropolitan region in which they play.[48])

In their 2001 study of the economic impact of five strikes and lockouts in the NFL and MLB, Coates and Humphreys found "that real income per capital in metropolitan areas did not fall during work stoppages in professional sports leagues, supporting the emerging consensus...that professional sports has no tangible economic impact on local economies."[49]

This is a "consistent conclusion" of those who have studied the matter, says Phillip A. Miller of the University of Missouri. Whether or not a city has a major-league team is not a significant economic factor within the SMSA (standard metropolitan statistical area) and, in fact, a franchise "could actually generate negative net benefits."[50]

Professor Miller, writing in the *Journal of Urban Affairs*, measured the effect on employment in the St. Louis area construction industry during the erection of the Trans World Dome and the Kiel Center. A construction boom is often proposed as one of the outstanding benefits of a new stadium, and one that justified the expenditure of public funds thereon. True, these jobs are of finite duration: once the palace of play is up, the jobs are done. But they pay well; they are visible; and those who perform them are usually members of influential labor unions. So there is considerable upside, in a politician's view, to "creating" such jobs from the public purse.

The Kiel Center was built primarily for the St. Louis Blues of the NHL. Ground was broken for the Kiel Center (renamed the Scottrade Center, after a local discount brokerage, in 2006) in March 1992; it opened in October 1994. The cost of $170 million was borne by a public–private partnership including the city of St. Louis (owner of the center) and local corporations, which guaranteed $98 million in construction loans.

The $280 million Trans World Dome, the contemporaneous construction project in St. Louis, was funded entirely by state and local governments. Built between May 1993 and October 1995 and later renamed the Edward Jones Dome, it had a very specific purpose: to provide a home for a new NFL franchise after the St. Louis Cardinals had decamped to Phoenix. It worked: the Los Angeles Rams moved to St. Louis for the 1995 season.

Professor Miller designed theoretical and econometric models to assess construction employment in the St. Louis SMSA before and during the period under which the Kiel Center and the Trans World Dome were under construction.

"We find no evidence," he writes, "that construction industry employment in the St. Louis SMSA was higher in the periods during which the Kiel Center and the Trans World domes were being constructed."[51]

Miller's conclusion? The statistical evidence, he says, "suggests that the levels of employment in the construction industry were neither higher nor lower during the construction of these stadia." Therefore, it seems that "construction on these projects merely substituted for other construction projects in this SMSA."[52]

The jobs supposedly created by sports arena and stadium construction "appear to be a direct result of purchases not made in other sectors of the economy." This is not new spending but reallocated spending. There is only a net economic benefit "if those workers employed on the stadium project would have been otherwise unemployed."[53] From Miller's study, this appears not to be the case. These construction jobs were not so much created as they were merely shifted from other projects that would have been pursued had the Kiel Center and Trans World Dome not been built; they resulted "in no new job creation in the construction industry."

After his extensive study of the St. Louis experience, Miller declares: "the net impact of stadium construction on construction employment and worker incomes is zero."[54]

Likewise, economist Ian Hudson of the University of Manitoba in Canada, in a paper for the *Journal of Urban Affairs*, created a model to isolate the effect on a city's economy of gaining or losing a professional sports franchise. His particular focus was on Winnipeg, which was engaged in an ultimately losing fight to retain the Jets of the NHL from being part of the (disastrous, in many ways) exodus of hockey teams from their native frozen grounds in the Great White North to the Sunbelt in the USA.

Winnipeg's civic leaders insisted that retaining the Jets was crucial to the city's image and prosperity. The Jets played their games in the Winnipeg Arena, which had been built in 1955–1956 with $2.5 million in loans from the city government of Winnipeg. The Arena became a storied hockey venue over the next decades, but by the mid-1990s it did not suit the owners of the Jets, who looked southward. City officials came up with a plan for a $100 million arena built with government funds; as a 1994 city position paper declared, "It has become apparent that North American communities with any prospect of becoming or remaining first tier will require a first-class sports, entertainment, and convention facility. Winnipeg now has that opportunity."

But the opportunity passed. The Jets moved to Phoenix. And Professor Hudson, writing from the province deserted by the Jets, concluded from his economic model that "the number of sports teams in a city has no statistical relationship to changes in employment."[55] Subsidization of a hockey arena makes no economic sense; the only possible argument therefore is that somehow the presence of a team has intangible psychic benefits in that it confers the "first tier" status desired by the makers of Winnipeg's argument for a new arena. (The reborn Jets of 2011–2012 played in the MTS Centre, built by the private True North Sports and Entertainment Limited, which owns the Jets.)

"Projected construction costs chronically are underestimated," concluded Dean V. Baim in *The Sports Stadium as a Municipal Investment*.[56] Most spectacular, of course, was the Superdome in New Orleans. And there is no evidence that government subsidies trickle down to the fan in the bleachers. As Baim notes, subsidies are, for the most part, "in the area of construction costs and property taxes on the stadium," which are "fixed costs, so little if any of the subsidy will be passed on to the fan." The income transfer is "perverse" — that is, from the middle-class taxpayer to the wealthy team owner and well-paid players.[57]

And then there are the subsidies seldom included in the dollar estimates: public construction of highways and off ramps and access roads, sweetheart rents, exemptions from taxes which are levied on other entertainment businesses.

"A precise rendering of the economic costs and benefits conferred by stadiums may make 'not in my backyard' a rational neighborhood response to those who seek a home for a professional sports team," argues Robert A. Baade.[58] The jobs and growth assertion adduced in support of public aid to such entities simply does not wash.

Bearing his torch of truth and light into the darkened corners in which stadium deals are made, Baade has time and again found and revealed that professional sports teams and the stadiums in which they play have no "significant impact on a region's economic growth." He has studied the matter from every angle, every spot on the field, and every vantage point from home plate to the end zone. Always he is forced to conclude that "the history of sports stadiums is indeed 'written in red ink.'"[59]

And that ink flows out of town, too. In "The Economic Impact of Sports Teams and Facilities," Roger G. Noll and Andrew Zimbalist point out that a majority of a team's gross revenue in all professional sports "goes to athletes" — so much for the players' union stereotype that these are exploited workers — while much of the rest "goes to owners, executives, on-field managers and coaches, and scouts."[60] Given that few of these persons reside full-time in the host city, a considerable portion of this revenue leaves the area.

The majority of spending (55–60%) by major league teams in the four primary professional sports is dedicated to salaries for players, who, by and large, seldom are full-time residents of the host city.[61] Most of this income is then spent or invested in cities other than those that had provided the direct subsidy of stadium or ballpark construction. The payroll monies that remain within the community are the relative pittance that go toward the usual part-time and seasonal employment provided by these teams: ushers, concession stand workers, and maintenance people.

Of course one person always, and without exception, benefits from a subsidized stadium: the owner of the franchise. A new venue provides "substantial short-term returns to team owners," writes Craig A. Depken II of the Department of Economics of the University of Texas at Arlington in the pages of *Public Finance and Management*.[62]

Professor Depken studied the effect of new stadiums on baseball team finances. His sample included the 15 ballparks that opened between 1990 and 2003, or right in the middle of the golden age of the single-purpose, retro-tinged ballpark era. Thirteen of the 15 were built with primarily public monies: Tampa Bay (100% public), Chicago White Sox (100%), Baltimore (96%), Arlington (Texas Rangers) (71%), Cleveland (88%), Denver (75%), Phoenix (Arizona Diamondbacks) (68%), Seattle (a devilish 66.66%), Houston (68%), Milwaukee (77.5%), Pittsburgh (100%), and Cincinnati (86.15%). The other three parks were built in Detroit (38% public financing), San Francisco (3.92%), and Atlanta (0%; it was built for the 1996 Olympics).

Previous studies have suggested that the novelty effect of a new stadium wears off in anywhere from three to five years, which is perhaps why owners quickly grow tired of aging stadiums and ask for new toys. Depken concluded from his study that, indeed, profits soar in the early years of a fresh ballpark: "The additional profit to a team owner averaged in excess of seven million dollars during each of the first five years of a new baseball stadium."[63] If you can get a public entity to build the stadium for you, as the vast majority of owners do, then public expenditures feed private wealth. Especially during the honeymoon.

But those profits don't seep down to the people in the neighborhood. A 2006 study of "The Impact of Stadium Announcements on Residential Property Values" in the Dallas-Fort Worth metroplex refuted the claim, made on occasion by stadium boosters, that residents of the city fortunate enough to host a major-league sports palace experience dramatic increases in property values. Economists Craig A. Depken II and Michael R. Ward of the University of Texas-Arlington and Carolyn A. Dehring of the Department of Insurance, Legal Studies, and Real Estate of the University of Georgia investigated the effect of five separate announcements by the Dallas Cowboys about sites that they were considering for their new stadium. Two concerned Fair Park, the neighborhood in which the Cowboys played in the Cotton Bowl before moving to suburban Irving in the early 1970s, and the other three concerned Arlington, which eventually "won" the dubious battle to bring the Cowboys to play in Cowboys Stadium beginning in 2009.

Their finding: the "aggregated expected city amenity effect [that is, the effect on property values in the affected area by such an announcement] and local sales tax burden associated with the proposal to build a publicly subsidized stadium for the Dallas Cowboys in Arlington, Texas, reduced residential property values."[64]

Cowboys Stadium was a good deal for Jerry Jones, no doubt — but not for his new neighbors.

The recent movement of stadiums back downtown, often as part of a general urban redevelopment plan, has failed to bear profitable fruit, as

Robert A. Baade notes in "Home Field Advantage? Does the Metropolis or Neighborhood Derive Benefit from a Professional Sports Stadium?" Baade, dean of economic studies of the sporting life, writes that the experience of American cities "strongly disputes the claim that professional sports teams and stadiums provide an economic boost for metropolises." After all, a "stadium does not resemble the corner grocery, where the owners live above the store."[65] Their wages flee the neighborhood, and often city, with the speed of a fleet halfback pulling away from a corpulent lineman.

Moreover, the newer stadiums are remarkably self-contained entities wherein one can not only see a game but buy a hat, a hot dog, a beer, sushi, an umbrella, a foam We're Number One finger, and anything else the fan's heart desires. Parking lots cut ballparks off from the surrounding neighborhoods, from shops selling t-shirts, from hot-dog vendors, and from small entrepreneurs. If you want a hot dog, you're going to pay $5 to the concessionaire inside the stadium. The days of street vendors and entrepreneurs hawking wares outside a stadium are as dead as old Tiger Stadium. This is especially true at Wrigley Field, whose constricted concourses limit the number of in-house vendors. To discourage enterprising folks from making a buck from selling food and t-shirts outside the ballpark, the Cubs are waging war on street vendors for "public safety" reasons — which is true if "public safety" is a synonym for "the Cubs' bottom line."

The infrequency with which football stadiums are used — ten games a year for an NFL team if one counts the two exhibition games; and at best, a playoff game or two on occasion — dampens any impact on the neighboring area. The practice in the 1960s and 1970s of locating such stadiums in the suburbs — think the New England Patriots in Foxboro, Massachusetts, and the Buffalo Bills in Orchard Park, New York — also limits their effect. As real-estate analyst John C. Melaniphy writes in "The Impact of Stadiums and Arenas," a pro football "facility in a remote location surrounded by a sea of parking with only ten games a year will have very little impact on the community other than image and pride."[66] (Melaniphy does note that one industry battens on the presence of a stadium, particularly in residential areas. Writing of the region around Wrigley Field in Chicago, he notes, "Illegal parkers during the baseball season are the towing company's bread and butter."[67] One man's parking ticket is another man's meal ticket…)

And it is not even as if moving a team to a new venue will give it a jolt. The owner who picks up his marbles and takes them to a new city may well find a newly enthusiastic fan base, but his team is likely to suffer on the field, court, or ice, as players struggle to assert a homefield advantage in a new place. Yet what of the team that remains in a city but moves to a different playing facility? Is it any better off? Richard Pollard, professor of Statistics at California Polytechnic State University at San Luis Obispo, discovered, "Home advantage during the first season in a new stadium after the move was significantly less than home advantage in final season in the old stadium."

Writing in the *Journal of Sports Sciences,* Pollard surveyed the growing literature on the factors that contribute to a winning team, which include the elements of the popularly believed "home advantage": crowd noise, travel, familiarity with the playing field or surface, even the intimidation factor against referees. These are "physical, sensory, and psychological," says Pollard.[68] Players at home understand the subtleties of lighting, the effects of wind and sun, and such idiosyncrasies as the way the bounce of a ball might be the product of local conditions. Presumably, the homefield advantage, while it would not disappear as the result of a move — the visiting team still suffers from the effects of travel and the referees will be susceptible to home-crowd pressure — will lessen, as players may take a considerable while to adjust to a new park or arena, even if it is located next door to the ruins of the old one.

To test this hypothesis, Pollard studied the 37 MLB, basketball, and hockey teams which switched venues (but not cities) between October 1987 and April 2001. He did not include football teams, since the number of games is too small to be relied upon, nor did he include teams that changed cities or switched to or from a dome to an open-air stadium.

The results were statistically significant. Of the 37 teams, 26 experienced a decreased home advantage in their new digs; ten experienced an increase, and one was a wash. The reduction was most dramatic in hockey, where the home ice advantage declined by a whopping 43.9% when a team skated into a new arena. Familiarity, it seems, breeds not contempt but sporting success. The price of a new ballpark or arena cannot be measured only in taxpayer dollars but, in most cases, losses on the field, court, or ice.[69]

Finally, if building ballparks for major league teams is a bad investment for taxpayers, so is publicly subsidizing minor-league parks. The authors collected in *Minor League Baseball and Local Economic Development* are in concord on this point. As editor Arthur T. Johnson writes, "In nearly all cases, the economic impact of a minor league team will be minimal, given attendance levels, payroll size, number of employees, media interest, and other factors."[70] Any claim by fervent boosters that a new taxpayer-purchased stadium for the local boys of summer will shower riches from the sky upon the infield should be regarded with extreme skepticism. A minor league baseball team may enrich a community in many ways, by either broadening entertainment options or providing a place for the generations to gather in good spirits to watch the great American game, but the economic benefits of the team will be, at best, modest.

"If new development is the goal," Johnson continues, "more is needed than building a stadium in the middle of a corn field and waiting for businesses to grow around it."[71] If you build it, they will come? No, more like if you build it, they will pay through the nose for it via taxes.

Minor case in point: Mark S. Rosentraub and David Swindell, writing in *Economic Development Quarterly*, analyzed the cost–benefit calculations that figured into the decision of the city leaders of Fort Wayne, Indiana, to refuse to offer "substantial inducements" in order to gain a minor league baseball team.[72] The situation was this: Fort Wayne, home to a quarter-million people and the second-largest city in Indiana, seemed a natural fit for the Midwest League, one of the premier Class A baseball leagues. In 1989, a Fort Wayne-based set of investors bought the Midwest League's Wausau Timbers, with an eye toward moving the team to Fort Wayne for the 1990 season. The only problem was that Fort Wayne lacked a baseball stadium that met professional standards. An existing facility could be upgraded, but at the cost of $1.95 million. The city offered the new owners a partial subsidy: a 15-year 6.48% loan for $1.2 million. The rest of the cost would have to be absorbed by the owners. The city would waive rental payments and maintain the stadium.

The owners were unable to raise the additional $750,000 or find private sources to guarantee the loan, so the deal collapsed. They sold the team. Rosentraub and Swindell ask: Did the city fathers and mothers make the right call? Should they have offered to "build the stadium for the Timbers and absorb all the costs of providing a playing facility"?[73]

Using data from other Midwest League teams, they reckoned the substitution effects and real growth impact that such teams have on the local economies. After all, dollars spent on baseball tickets and hot dogs and beer may simply have been shifted from other expenditures within a local economy — say, on economics texts or strip joints or rap concerts or martini bars. Cracks Vanderbilt University economist John Siegfried: "What are people going to do with their money if they don't spend it on the Red Sox, flush it down the toilet? No, they'll spend it on something else: books, maybe, or bowling, things that Boston would benefit just as much from."[74]

For Fort Wayne, the activity stirred by the presence of a minor league team would not have been worth subsidizing the cost of the stadium, found the authors. The $1.95 million cost to the city would have amounted to "a very large fiscal loss" that would not have been made up by projected economic growth. In sum, "even optimistic measures of their income would not sustain the costs of operation of team, renovation of the stadium, and any return on the investment." This is true throughout minor league baseball, the authors note. On "purely economic grounds," subsidizing minor league stadiums is a bad deal. So "Fort Wayne's decision not to finance the stadium was correct."[75]

But there is a footnote: Fort Wayne's city leaders, having made the economically wise decision in 1989, turned around in 1993 and built Memorial Stadium for the Midwest League's Fort Wayne Wizards, who had relocated from Kenosha, Wisconsin. In 2009, Memorial Stadium was demolished, and the Wizards — now renamed the Fort Wayne TinCaps, after Johnny Appleseed's headgear — moved into Parkview Field, a $30 million facility, $25 million of which the city of Fort Wayne paid in order to provide a home for the TinCaps, who are owned by an Atlanta outfit called Hardball Capital. Drew Stone, writing in the Fort Wayne *News-Sentinel*, notes that the $2 million demolition of Memorial Stadium "hasn't pleased residents who wonder why a 16-year-old facility should be replaced and why the city is spending tens of millions of dollars" on the new stadium, which is part of a downtown development project which also includes condominiums, a hotel, and a parking garage.[76]

A good question — and one that Fort Wayne's city leaders answered correctly back in the 1980s.

Tax That Fellow Behind the Tree!

In recent years, the old reliable of bond issuances has been joined by the allegedly painless hotel tax as the most common way of publicly subsidizing stadium construction. The hotel tax, it is asserted, doesn't really hurt locals, since those being taxed are presumably out-of-towners who don't even know they're being fleeced. It is consistent with the late Louisiana Senator Russell Long's ditty about the preferred method of revenue raising: "Don't tax you/ Don't tax me/Tax that fellow behind the tree."

But as Victor A. Matheson of the College of the Holy Cross and Robert A. Baade of Lake Forest College conclude from their examination of the financing methods of 20 NFL stadiums built between 1992 and 2006, while the prospect of taxing outsiders via hotel and rental car levies may have a surface appeal, "the benefits of these taxes are not nearly so clear."[77] A substantial percentage of car renters are locals, for one thing. And is there evidence, ask Matheson and Baade, that those who rent cars after experiencing accidents are disproportionately sports fans? Moreover, a tax on transients only shifts the burden to outsiders if the city in question is the only city imposing such a tax. But as such taxes become widespread in major-league cities, then traveling football fans are subsidizing via hotel taxes other teams, if not their own.

Nevertheless, the frequency of cities imposing rental car taxes to fund stadiums tripled from the 1990s through the first decades of the twenty-first century, and hotel taxes are now so common as to be virtually the default method of subsidization.[78]

Matheson and Baade say that the "least inequitable" method by which a professional team might be subsidized would consist of a combination of fan consumption taxes, personal seat licenses, ticket surcharges, and perhaps "a small tax increase spread over a city's population."[79] In other words, require those who consume the product to pay for it. But given the fact that those consumers are usually the loudest voices within a community favoring subsidy, they are unlikely to advocate taxing themselves for their pleasures. Far better to tax the non-fan.

Government stadium builders have a multitude of taxes in their bag of tricks. Besides hotel guests, they tax car rentals, restaurant food, alcohol and tobacco, the wages of professional players, casinos, cable television bills...the list is limited only by the imagination and the brazenness of the taxers — and, on those occasions when the populace has a say, by the willingness of the voters.

The greater the public say, the less likely a stadium will rely on sales taxes for revenue. For the sales tax affects any consumer in the designated city, county, or state. Far better to lay the tax upon a wayfaring stranger, an innocent (or not-so-innocent) who happens to book a weekend room at the downtown Holiday Inn in order to enjoy a town's nightlife or museums or to transact business. And if he happens to rent a car in order to see the sights or make his meeting, so much the better. The best feature of the taxed traveler is this: he doesn't vote in local elections.

Mayya M. Komisarchik and Aju J. Fenn note that while almost 60% of referenda on using public monies to construct stadiums or arenas were defeated in the years 1974–1996, a whopping 84% (21 of 25) were passed in the decade of 1995–2005. Moreover, although "[r]eferendum propositions for PNC Park [Pittsburgh], Safeco Field [Seattle], Soldier Field renovations [Chicago], and a Charlotte Arts Arena package were defeated...in each case a special legislative session, city council vote or independent agreement between the team franchise and the municipality circumvented existing obstacles and provided for stadium construction."[80] If at first they don't succeed, they change the rules of the game.

Anti-stadium subsidy coalitions range "from social-justice activists who advocate spending public money on schools instead of stadiums to foes of any government spending whatsoever," write Neil deMause and Joanna Cagan.[81] They may disagree on what they stand for, but they know what they *don't* stand for: the coerced transfer of wealth from those who pay the taxes, often on hotel stays and car rentals and general sales, to the corporate-welfare-seeking sports entities that benefit from such subsidies.

Yet even the doughtiest of anti-stadium activists and organizations are often overwhelmed. Chimes in Veronica Z. Kalich of Baldwin-Wallace College, "In no other industry is there such a clear case of the ineffectiveness of taxpayer resistance."[82] It is not entirely hopeless — resistance is not quite futile — but the odds are stacked against the underdog taxpayers.

"I believe the citizens should have a say in this issue," declared Chandler, Arizona, mayor Jay Tibshraeny in 1995 in reference to a referendum in which Chandlerites were to vote on renovations to a spring training facility for the Milwaukee Brewers. The mayor helpfully explained, "If the voters pass this, we'll move forward. If the voters don't pass this, we'll still move forward."[83] Heads he wins, tails you lose.

In his examination of direct democracy and stadium funding, Rodney Fort, Professor of Economics at Washington State University, found that in the 29 yes–no popular votes he studied, stretching from 1974 to 1996 and including the cities of Arlington (TX), Baltimore, Charlotte, Cincinnati, Cleveland, Denver, Detroit, Nashville, San Antonio, San Francisco, Chicago, Colorado Springs, Durham, Miami, Milwaukee, (state of) New Jersey, Oklahoma City, San Jose, (county of) Santa Clara, and Seattle, a few general rules seemed to apply. "Excise taxes were always approved, whereas utility taxes and a sports lottery were never approved."[84]

The votes were usually close; more than half had yes votes in the range of 45–55%. But it is tough to fight the combined forces of city hall, the Chamber of Commerce, and the local pro sports celebrities — especially when you are routinely outspent by 40 or 50 to 1.

Of course there are alternatives to taxes, and some do not even involve the owner digging into his personal wealth. The "personal seat license," or PSL, is a creative way to require ticket holders themselves to pay part of the cost of the venue. By paying the PSL fee, which ranges from a few hundred dollars to several thousand dollars, the fan reserves a particular seat for a specified number of seasons. He or she then must also purchase a season ticket. So the PSL is simply an additional payment on top of the cost of season tickets; in the case of subsidized stadiums, it adds insult to injury, though at least it burdens those who actually use the stadium.

New parks have luxury suites, restaurants, bars, apartments and hotel rooms, concessions that are something more than hot dogs and a coke, parking, club boxes, and other accouterments that would astonish a Yankees fan of the 1920s — or even the 1970s. These ballparks employ new construction technology to "maximize opportunities for revenue generation from luxury suites, club boxes, concessions, catering, signage, parking, advertising, and theme activities," writes Andrew Zimbalist.[85] This has enabled smaller cities in the South — Jacksonville, Charlotte, Nashville, Memphis — to enter the big leagues, in some cases luring teams from much larger cities (the Tennessee Titans from the Houston Oilers) or outcompeting bigger cities for new franchises. But as Zimbalist notes, this expansion of the number of economically viable cities also expands the "imbalance between supply of and demand for sports franchises."[86] As a result, team owners have a deeper pool of team-less cities to which they can threaten to move if the current host city doesn't cough up enough lucre to keep them moored.

Surprisingly, as Komisarchik and J. Fenn explain in detail, stadium and arena capacities have not grown in proportion to construction costs. Comparing pre- and post-1995 venues, they write, "Major League Baseball stadiums built before 1995 could seat an average of 45,368 people, while more recent ballparks seat an average of 44,419 people. The average capacity of older NBA arenas is 19,652 patrons, while new NBA arenas seat 19,159. Old NHL arenas have an average capacity of 18,244, and modern of 18,381. Finally, seasoned football fields hold 71,162, whereas new stadiums house only 70,563 fans." While projected capacities for sites on the drawing board are somewhat higher for the NFL, the obvious fact is that stadiums and arenas are not getting bigger, although they are getting more expensive. Komisarchik and Fenn calculate that in real dollars, the average construction cost of an MLB stadium was $187 million before 1995 and $389 million from 1995 to 2009; comparable figures for the NFL are $182 million pre-1995 and $365 million post-1995; and for an NBA arena, construction costs rocketed from $177 million real dollars before 1995 to $316 million post-1995. The anomaly in the study was the NHL, whose average arena cost in real dollars actually fell from $217 million pre-1995 to $205 million post-1995.[87]

The slightly smaller capacities may be explained in part by the emphasis on luxury seating over standard squeeze-your-butt-into-the-chair seating, and, in baseball, by a desire to capture the retro feel of intimate parks like Fenway.

In some cases, authentic old parks were demolished to make room for faux old parks. Tiger Stadium and Comiskey Park, venerable redoubts of baseball's heyday, were both consigned to the junk heap even though ardent fans tried to save them. After losing a battle to save his beloved park, Frank Rashid of the Tiger Stadium Fan Club learned this lesson, as he told deMause and Cagan: "The local politicians, particularly the mayor and the county executive, know that they get far more mileage out of having a big new project than out of a renovation. They have the ability to say who gets the contracts, whose land is used, which developers are employed, which bond attorneys do it — and all of those people are the people who contribute to their campaign war chests."[88]

Taking up the cudgel against nostalgia, Allen R. Sanderson, a critic of blatant giveaways, suggests that ballparks and arenas "constructed before the advent of television, jet travel, air conditioning, and computers" might be "simply outmoded, worn out." They are often "dank," and marred by "narrow aisles, obstructed views, and limited restroom and concession space."[89] All true, though there remains the question of who should pay for upgrades, renovations, or even replacement parks. If Fenway, Tiger Stadium, Comiskey Park, and other venerable ballparks were built, for the most part, with private funds, why should the (largely characterless) parks that have or have been proposed to supplant them — and perform exactly the same function — be the responsibility of state, county, and municipal taxpayers? Narrow aisles are no doubt an inconvenience, but so is a widened tax base, and those negotiating those narrow aisles are doing so voluntarily, unlike those upon whom stadium taxes are levied.

Economic Impact Studies — Or, High Fiction with Multipliers

The economic impact of a team or stadium on a community is usually measured in three ways: by direct expenditures, indirect expenditures, and, nebulously, accrued "psychic" benefits.

The first of these consists of the most visible spending due to a team and its stadium: the money people spend at restaurants, bars, hotels, rental car agencies, and the like as they come into town for a game.[90]

Indirect expenditures are, in a sense, further branches of this spending: the restaurant owner spends the monies received from fans on a new car, the car dealer spends the funds thereby received on remodeling his kitchen, and so on down the line. This is, as you might imagine, a field fraught with potential for distortions, and in fact the "multiplier effect," which we shall consider in a moment, is more elastic than a politician's definition of truth.

Finally, there are the "psychic" benefits, almost impossible to quantify, which consist of the sense of pride or satisfaction or pleasure one gets from having a team in one's town. These, as we shall see, are a booster's delight — easy to exaggerate because they are so hard to quantify.

The Economic Impact Study is a crooked arrow found in the quiver of every city and team official trying to persuade taxpayers or their representatives of the wisdom of shelling out public monies to a pro stadium. Yet these are notoriously baseless, even nonsensical, their outrageous estimates ginned up by the misuse of the "multiplier effect" of government spending. The methodological errors rampant in such studies would gag a mediocre Econ 101 student. Yet "any consultant who dared provide an unfavorable report would probably be fired or unable to get new contracts in the future," observes Steven A. Riess.[91]

Briefly, the multiplier effect is based on the truism that money circulates; if, say, the government pays a construction company to build a stadium, the construction workers, for instance, may spend their paychecks on pizza or car payments or church donations or strip clubs or whatever. The recipients of this money will in turn spend it on other goods and services. Thus the "multiplier effect." But even a middling Econ 101 student knows that there is also something called an "opportunity cost": that is, the cost of foregoing the next-best option when making a decision. Money transferred in the first place from the city government to the construction worker would have followed a different path had the stadium not been built; it would not simply have vanished into the thin air. These alternative uses — what might have been; or the best other use of the money — are the opportunity costs. The money government spent to build XYZ Stadium could have been spent on mass transit, or hospitals, or schools — or it could have been given back to the taxpayers (or never taken from them in the first place). Joseph L. Bast asks, how "many private investments...don't get made because the money was confiscated to build or operate a stadium"?[92]

The problem is opportunity costs are overlooked — whether naively or cynically — by those who draw up rosy economic impact scenarios. As Marquette University economist William J. Hunter wrote in a paper on the misuse of economic impact studies, "The use of a multiplier to compute the total impact of a public works project on the local economy guarantees that estimates of community benefits will always exceed estimates of community costs. . .Moreover, the larger the project undertaken, the greater the growth in community income."[93]

Hunter calls this the "Taj Mahal Syndrome." Public projects will always be found to be "worth it," and the bigger and more expensive the project, the more "worth it" it is. The implication is that spending large sums of taxpayer money on public projects is *always* a beneficial expenditure. Sluggish economy? Build an 80,000 seat stadium! Spending your way to prosperity was never so easy.

Multipliers in the most roseately unrealistic impact studies are as high as 2.5, meaning that two-and-a-half jobs are projected to be created for every job initially created. No competent economist takes such a number seriously.

Moreover, multipliers ignore the fact that the primary recipients of salaries from the teams that play in these new stadiums — the players — usually live (and spend and invest) outside the area. On the other hand, the bar and bowling alley owners and employees whom the sports fan might patronize in the absence of the professional team almost always do live and spend in the area. The local multiplier is considerably higher when one spends money on local entertainment rather than on major-league sports. The multiplier is — or should be — substantially lower when a sizeable portion of monies spent on pro sports goes to athletes and owners who live outside the area.

So in this sense, the taxpayers are actually subsidizing the flight of capital. And as we have previously seen, the trend today is for stadium operators to offer a wide range of culinary and subculinary (e.g., stadium hot dogs and seven dollar cups of Budweiser) choices to fans, who thus spend their entertainment dollar within the stadium rather than at neighborhood restaurants and bars. As one analyst writes, "The new generation of fully-loaded facilities is likely to capture much of the spending that used to occur in nearby restaurants, bars and sports merchandise stores."[94] (Concessions companies, unlike local taverns, are often based in distant cities as well.) Owners are happy to have local tavern owners show up at city council meetings to speak on behalf of increased sales and hotel and rental car taxes, but their sense of comradeship with these small-business owners does not extend so far as to cede to them responsibility for fulfilling the thirst of game-day fans.

Economic impact studies trumpet the gross benefits of a team's presence without considering — indeed, while studiously ignoring — benefits foregone. That is, the ways in which people would have spent their money on other entertainments or services had the sporting option not been available. A fan who shells out 100 dollars for tickets and parking and beer at the football game might have spent that 100 dollars on a night at a steakhouse or a blues concert. The alternative spending choices of state and local governments, consumers, fans, taxpayers, and patrons would be most inconvenient pieces of these equations, and so they are left out of the discussion. Yet as Robert A. Baade insists, "The alternative of not taxing the money from citizens in the first place should always be included in this evaluation." And as Baade adds, "Public officials must inevitably ask, Which is more valuable — more money for schools, roads, airports, police, tax reduction, or stadiums?"[95]

Economic impact studies paint beyond-rosy scenarios of the benefits of publicly subsidized stadiums: the construction jobs, the restaurant and hotel spillover, the increased tax revenues for the host city and county — why, the benefits flow like lemonade springs through the big rock candy mountain! The academic literature on the subject reaches quite the opposite conclusion, but still the economic impact studies are authorized and published and cited as gospel by boosters. In recent years, pro-stadium impact studies have claimed that well over 4,000 "permanent jobs" would be created if stadiums were built in Cincinnati, Kansas City, and Phoenix, for instance. In fact, say Federal Reserve economists Jordan Rappaport and Chad Wilkerson, who are not unsympathetic to stadium subsidies, "the net number of jobs created from hosting a professional sports team is quite low. It is almost certainly less than 1,000 and likely to be much closer to zero."[96]

John Siegfried and Andrew Zimbalist point out that in addition to the 100 or so front-office people employed by the average professional team, a football team will typically hire 1,000–1,500 persons for day-of-game work in part-time, low-wage positions. At four hours/day for ten home games a year, this adds up to a grand total of "20 to 30 full-time, year-round jobs."[97] It is tough for the Chamber of Commerce to fly that stat on its mast.

As Kevin J. Delany and Rick Eckstein of the Departments of Sociology of Temple University and Villanova University, respectively, conclude in their paper "The Devil Is in the Details: Neutralizing Critical Studies of Publicly Constructed Stadiums," pro-stadium forces have adopted an array of strategies to deal with critical studies and with academic refutations of rosy scenarios. To wit, "pro stadium elites have ignored the [critical] studies, criticized them without competing evidence, commissioned contradictory studies, or shifted the debate to non-measurable endpoints."[98]

Delany and Eckstein studied nine cities in which subsidy battles had taken place: Hartford, Cleveland, Cincinnati, Philadelphia, Pittsburgh, Minneapolis, San Diego, Phoenix, and Denver. The pro-stadium strategies can be boiled down as such: Ignore, Attack, Change the Subject.

In the Ignore category, the authors cite stadium advocates who go about blithely ignorant of the voluminous evidence from the other side. They do not acknowledge that substitution effects exist, for instance, and they frequently hold up Baltimore and Cleveland as examples of cities that were turned around by subsidized stadiums — despite the lack of empirical evidence. Camden Yards of Baltimore acts as a kind of talisman, as a magic charm that screams "Revitalization" even though, as the authors note, Baltimore's Inner Harbor revitalization preceded the construction of the new baseball and football stadiums. As for Cleveland's Jacobs Field (now Progressive Field), the authors asked "someone central to building Jacobs Field" for the research upon which he based his claims for its resuscitative effects, and he replied: "It was not studied. I just knew it. I would not waste $100,000 for someone to study this. It's the right place to build."[99]

He just knew it. Cuyahoga County sunk $84 million, raised by alcohol and cigarette taxes on its citizens, into Jacobs Field, or about half of the total cost, and yet $100,000 was too much to spend to determine whether or not the $84 million public investment made sense.

But then, even if such studies were made, any contradictory evidence would have been buried or condemned. That, after all, is how stadium boosters deal with the shelves full of academic research on the topic. As Delaney and Eckstein write, when ignoring hostile studies doesn't do the trick, the practice is to mock and ridicule them. Why, the professors whose names are attached to these papers probably can't even hit a curve ball! They "just don't understand the nuances of stadium generated economic development," and besides, they've "probably never even been to a new ballpark."[100] They just don't get it. And what are those eggheads doing sticking their noses into sports anyway?

Just in case any taxpayers get the bright idea to rely upon peer-reviewed research in organizing opposition to stadium giveaways, pro-subsidy advocates maintain the option of commissioning their own economic impact studies. These tend to be drawn up by accounting firms rather than professors of economics, and in their reliance on questionnaires and surveys and speculation and "sloppy methodology," they are "fantasy documents" which promise the moon and stars and an El Dorado-style prosperity if only the city council or county legislature or the voters will approve the requested subsidy. He who commissions the study calls the tune.

These fantasy documents look backward as well as forward. Retrospective studies, "dressed in methodological regalia," attempt to show that publicly constructed stadiums have delivered on their promises.[101] The problem with such studies, say Delaney and Eckstein, is that they are the merest burlesques of real studies. Typically, as was the case with a 1997 Arthur Anderson LLP study of "The Impact of a Ballpark in Central City Philadelphia," the "researchers" rely on fan surveys that may oversample out-of-town fans (who spend considerably more on non-ticket purchases than others) and undersample locals, children, and others who spend less.[102] The result is an exaggerated claim of per-fan spending.

They also greatly exaggerate the number of out-of-town fans who come to games. A 1997 survey of the literature set the range of such fans at between 5% and 20%, and yet, for instance, one economic impact study in support of a new Fenway Park claimed, absurdly, that 35% of fans at Red Sox games were from out-of-state. ("The methodology employed to arrive at the 35% estimate is not described," state Siegfried and Zimbalist archly.)[103] Moreover, evidence suggests that the majority of out-of-area fans in the stands do not come to the host city solely for the purpose of attending a game. So if they were not at Fenway, say, they would be spending those dollars on some other Boston attraction.

Additionally, such surveys ignore the substitution effect. The possibility that a fan might have spent his baseball ticket and hot dog money on something else seems entirely to have escaped the attention of the economic impact study authors. But then that's the point: shoot at everything that flies and claim everything that falls is the economic development mantra.

In researching the ways in which pro-stadium forces neutralize critical studies, sociologists Delaney and Eckstein write, "In all of the nine cities we studied [Hartford, Cleveland, Cincinnati, Philadelphia, Pittsburgh, Minneapolis, San Diego, Phoenix, and Denver], the major local newspaper editorially championed spending public dollars on private stadiums…Editorials supporting the new stadiums would often parrot the fantasy documents" purporting to be economic impact studies put out as propaganda by the pro-subsidy side. Newspapers tend to be pro-subsidy, since their readership and circulation are tied, in critical ways, to the presence of major league teams.

Interestingly, there was one significant exception to this rule, and it has implications for other battles in this long war. To quote from Delaney and Eckstein at length:

> The potential influence of a less sycophantic media was apparent in Pittsburgh, where the city's "second" mainstream newspaper (*The Pittsburgh Tribune*) was as critical of publicly financed stadiums as the "first" paper (*The Pittsburgh Post-Gazette*) was supportive. In addition, the libertarian publisher of this paper, Richard Scaife, owned his own think tank, the Allegheny Institute on Public Policy, which generated scores of position papers critical of publicly financed stadiums. These reports, of course, were picked up and legitimated by the *Tribune* where they had a very important impact on Pittsburgh's new stadium initiative. In fact, in almost unprecedented fashion, stadium advocates were crushed in a referendum seeking to raise local taxes for these two new ballparks and had to completely rethink their strategy (which was ultimately successful). The *Tribune*'s hostility certainly contributed to this surprising electoral defeat. Perhaps other referendums in other cities would have unfolded differently had the local media not been economic and political bedfellows of pro-stadium forces.[104]

Ominously, "an increasingly popular strategy" for neutralizing the academic literature on stadium subsidies is to "argue that it is not really about economics anyway." Mere dollars and cents can't possibly measure the value of a major league team to a city. Perhaps dimly perceiving that a veritable Everest of evidence is against them, the advocates of subsidy shift the focus of their arguments to such warm and fuzzy and immeasurable benefits as "solidarity, happiness, or family togetherness."[105] It is almost enough to make a credit card commercial. Cost of New Stadium to Taxpayers: $140 million. Benefit of Paying $300 for tickets and parking for a family of four: Priceless.

Except it is not priceless. And the proclaimed blessings of having a subsidized stadium are so dubious as to defy description. Delaney and Eckstein ask, "How could a systematic academic study possibly refute this piece of 'evidence' we were offered to support Cleveland's use of public dollars for private stadiums: the bike messengers around town were smiling more since the new ballpark opened."[106]

Ridiculous as such assertions may sound, they are the most tangible evidence that subsidy-boosters really have, and Delaney and Eckstein predict that the "strategic shift to emphasizing noneconomic issues like community self-esteem and community collective conscience" will only increase. Interestingly, most of the business leaders the authors spoke to in the nine cities they studied were not under the illusion that "new stadiums would provide community economic windfalls"; you'd have to be a politician to believe that. But they often saw it as their civic duty to support the imposition of new taxes to fund such projects — though in some cases, the fact that they were "avid sports fans" drove their enthusiasm. As the authors conclude, "There are many powerful people who remain supportive of using public money for private stadiums, and they will keep trying to come up with new ways to avoid dealing with the devilish details."[107] The hope, for those who oppose these powerful people, lies in the aphorism attributed to Mahatma Gandhi: "First they ignore you. Then they laugh at you. Then they fight you. Then you win."

But Gandhi was fighting British imperialists, not men chasing public handouts of hundreds of millions of dollars. When that sort of money is at stake, never underestimate the cleverness of the subsidy-seeker. From the desperate scramble for dole dollars comes the nebulous and very useful concept of "psychic benefits."

The admen who create the Mastercard commercials would say that the value of, say, boasting that the Steelers are from your hometown of Pittsburgh, or that the reigning World Series champs play in your county, is "priceless" — a helpfully abstract concept for those who wish to convince policymakers of the need for a larger cut of a hotel or sales tax. A team is said to be a symbol, a commercial ambassador, for its host city. Novelist James Michener, he of the doorstop-sized bestsellers, said in 1976: "a city needs a big public stadium because that's one of the things that distinguishes a city. I would not elect to live in a city that did not have a spacious public building in which to play games, and as a taxpayer I would be willing to have the city use my dollars to help build such a stadium, if that were necessary. I am therefore unequivocally in support of public stadiums."[108]

One owner put it this way: "Tonight, on every single television and radio station in the USA, Seattle will be mentioned because of the Mariners' game, and tomorrow night and the next night and so on. You'd pay millions in public relations fees for that." Yet as Siegfried and Zimbalist ask, "Do people view Charlotte, Jacksonville and Nashville to be big-time locations and Los Angeles an also-ran place because the former have NFL teams and the latter does not?"[109]

Obviously not, but owners and league officials play upon the fear of boosters of looking bush league. No catastrophe is quite so great as having the "local" band of sports mercenaries skip town. Gene Budig, American League president, said in 1995: "No community today wants to lose a franchise. It would send the wrong message to business and industry that might have an interest in it."[110] Added the head of a task force responsible for the horrid Hubert Humphrey Metrodome in Minneapolis: "It is almost worse for a city's image to lose a major league team than to have never had one at all."[111] Examples spring to mind — is it better to be Newark, New Jersey, which never had a major league team, or Los Angeles, which lost the Raiders and Rams in short order? — but they do not quite make the speaker's point.

(This is not to say that people are not deeply attached to their home teams, and that the departure of those teams doesn't exact a real cost. Charles C. Euchner, in *Playing the Field: Why Sports Teams Move and Cities Fight to Keep Them,* notes that "Psychologists have compared the loss of a sports franchise to the trauma experienced at the death of a loved one."[112] The same is true for those who love proud old ballparks — Tiger Stadium, Comiskey Park — which fall to the wrecking ball.)

Civic pride, as one subsidy advocate writes, "cannot be captured solely through ticket sales or the purchase of logo merchandise."[113] True. But these at least are measurable; by contrast, the psychic benefits of a person sharing a city with a professional sports team are so unquantifiable as to make the concept of *love* seem like it is reducible to a mere number.

If subsidies are to be defended, it must be on less measurable grounds — community pride and the like. There simply are no strong economic justifications of such public expenditures. And yet the urge overwhelms to wipe away even these fancy rationalizations for pocket-picking. Economic development, city pride: phooey. Much of this is just old-fashioned pork flavored with bribery and payoffs.

Still, pursuers of tax dollars are nothing if not creatively argumentative. For there is also, floating out there in the nether world between baseless assertion and wishful thinking, the claim that employers base their decisions on where to locate factories or offices or corporate headquarters in part on the presence or absence of professional sports teams. This claim adds heft to the pro-subsidy argument, making it appear as if hordes of free-agent businesses are ready to descend upon a city and its suburbs if only that city will give the team what it wants. Yet as Robert A. Baade writes, "no evidence exists to suggest that professional sports is an important factor in business location decisions."[114] There is evidence, however, that the local tax climate has a role in such decisions — but the tax climate most conducive to positive location decisions is one in which levies are lower rather than higher.

Contingent valuation method, or CVM, is a somewhat controversial means by which marketers, government agencies, and some economists attempt to determine the market value of nonmarket resources or goods (often environmental in nature, for instance a beautiful view or pristine wilderness) by asking participants in a survey what they would pay for — to give an example or two — a week without clouds or the presence of an all-Shakespeare theater in their town. Many economists insist that people's preferences are more accurately determined via market transactions than by off-the-cuff responses to hypothetical questions, yet believers in the efficacy of CVM have contributed to the literature on the psychic benefits of sports stadiums.

In 1997, economists Bruce K. Johnson of Centre College (Kentucky) and John C. Whitehead of East Carolina University applied the CVM to assess the value of a pair of projects proposed for Lexington, Kentucky: a basketball arena for the University of Kentucky Wildcats and a minor league baseball stadium that would be used to attract a team to Lexington, then the largest city in America without its own pro baseball team.

In the former case, although the proposed arena would have been privately funded, it would have deprived the publicly funded Rupp Arena, built in 1976 by Fayette County, of its major tenant. Given that taxpayers would not retire the arena's debt until 2016, the departure of the Wildcats would have increased the burden on county taxpayers.

In the latter case, a local developer told county officials that he could bring a Class AA Southern League baseball team to Lexington if only the county would provide a $10–$12 million park for the boys to play in.

Johnson and Whitehead surveyed a random sample of Lexington households, asking a series of questions about the extent of the respondent's interest in sports and culminating with the question, "Would you be willing to pay $x per year out of your own household budget in higher taxes to help pay" for a new arena/new baseball stadium?

More than two-thirds of those surveyed replied that they would not be willing to pay any amount in higher taxes for a new arena for the UK Wildcats, while almost as many (63.3%) opposed any hike in taxes to pay for a baseball stadium. More than one-third of those who were willing to be taxed for baseball cited the alleged economic impact, a sign of "stadium illusion," comment Johnson and Whitehead, referring to the vast body of evidence that there are no such benefits.[115]

The denouement in Lexington turned out to be virtually a best-case scenario. The Wildcats remained at Rupp Arena, and after local officials refused to build the baseball park, a group of private investors bought a franchise (the Lexington Legends) in the Class A South Atlantic League and financed — without taxpayer subvention — the $13.5 million Applebee's Park, which opened in 2001 and where the team entertains fans every summer — on those fans' own dime. (For the 2011 season, naming rights have been purchased by Whitaker Bank.)

Using the "CVM approach," in 2000, Johnson, Whitehead, and Peter A. Groothuis asked Pittsburgh area residents, "What is the most you would be willing to pay out of your own household budget each year in higher city taxes to keep the Penguins in Pittsburgh?" The answer was roughly 83 cents to $2.30 per person per year. With a 6% interest rate over 30 years, this translates to a subsidy of $26.9 million to $74.7 million — even at the higher number, this does not equal the average hockey arena subsidy in 2000.[116]

Taking the concept of "psychic income" about as deeply as it can go — making it the anchor of subsidy justification — is John Crompton of the Department of Recreation, Park and Tourism Sciences at Texas A&M University. Professor Crompton, writing in the *Journal of Sport Management,* sees the psychic-income leg as the only viable support for the expenditure of public funds on sports stadiums and arenas. Crompton acknowledges that the preponderance of studies over the last two decades has demonstrated that there is "no statistical relationship between sport facility construction and economic development."[117] Any defense of subsidies based on their alleged economic benefits is specious, if not an outright lie. They simply cannot be defended on economic grounds, unless one is willing to play fast and loose with the facts — perhaps, as we often see, with the encouragement of a fat commission from a local Chamber of Commerce.

But professional teams, Crompton says, can help cement a sense of community. They set a mood, an atmosphere, an ambiance. They feed pride. People identify with the team, even draw a sense of well-being from its successes (while they mourn its failures). "Emotional involvement" with a professional sports team, he claims, "transposes some people from the dreary routines of their lives to a mode of escapism that enables them to identify with a team, personalize its success, and feel better about themselves. Life is about experiences, and sports teams help create them — albeit vicariously."[118] (The hit to the taxpayers' wallet, it should be noted, is certainly not vicarious. More like vicious.)

Nevertheless, redefining the benefits of publicly financed parks and arenas as psychic rather than economic is the "new frontier" in subsidy apologias, says Crompton. Such a redefinition, he says, would permit defenders of stadium subsidies to fight upon higher (indeed, almost ethereal) ground. They needn't sully their integrity by pitching bogus studies. He writes: "By shamelessly using flawed economic rationales to justify subsidies of major facilities, elected officials and other community elites are characterized as untrustworthy, manipulative, charlatans with an agenda to sell. If a new psychic income paradigm for justification and scientific measurement is used to appraise the value of the psychic income and, hence, the subsidy invested, then these proponents could reposition themselves as responsible keepers of the public trust."[119]

This is a tall order, though not so tall as making straight-faced assertions that, say, giving $325 million to Jerry Jones is good for the Dallas economy. The great thing about psychic income is that it is so immeasurable that it is hard to empirically demolish. Sponsor a CVM study, ask the residents of a town without pro football how much they would pay to have a team, and since their answers don't actually cost anybody anything they will reply, more likely than not, with an inflated estimate of the value they place upon securing such a team. This may indeed be the wave of the future: instead of commissioning economic impact studies, a Chamber of Commerce in conjunction with the owner of a stadium-seeking team will commission psychic impact studies. Don't be shocked if the results are exactly what those who commission these studies want them to be.

(Anti)-Trusting in Expansion

Roger Noll, Stanford University economist, explains one effect of baseball's antitrust exemption: "Major league baseball should have 40 or 50 teams, not 26 or 28 [today 30]. Then you wouldn't have to give away hundreds of millions of dollars to get one to relocate."[120]

But of course that is exactly why the majors do not expand to 40 or 50 teams: they want to keep a pool of franchise-less cities begging outside the gates, offering bejeweled parks to footloose owners. And yet it is far from clear that stripping baseball of its antitrust exemption would have the desired effect. There is a better way to restore stadium sanity.

Baseball's antitrust exemption was established by the U.S. Supreme Court in *Federal Baseball Club of Baltimore, Inc. v. National League of Professional Baseball Clubs, et al.* (1922), since refined. The unanimous decision found, in the words of Oliver Wendell Holmes, that baseball was not "interstate commerce," and therefore not subject to federal regulation. But the Court also denied that baseball was "trade or commerce in the commonly-accepted use of those words," which somewhat unusual perception freed the industry from the oversight of antitrust laws. (Coincidentally, a lower court had delayed this case by a year, thereby pressuring the upstart Federal League to settle with MLB. The lower court judge was Kenesaw Mountain Landis, district court judge of the Northern District of Illinois. In gratitude, MLB named Landis its first commissioner in 1920.)

The court's decision, written by Holmes, asserted that although the "organized business of giving exhibitions of baseball between clubs representing different cities" frequently requires those clubs to travel from state to state, this travel does not constitute "interstate commerce," and therefore does not place the business of professional baseball under federal antitrust law. Putting aside the question of the wisdom and justice of antitrust laws, for Holmes and the Court to have decided, unanimously, that baseball did *not* have the characteristics of interstate commerce is remarkable. When the New York Giants ventured to Pittsburgh to play the Pirates, were they not crossing state lines for the purpose of engaging in a commercial act?

Holmes somewhat ingeniously suggested that because of the interdependent nature of the league, in which each team relied upon the others to furnish a field, to fill the stands, etc., the "personality" of each club "is actually projected over state lines and becomes mingled with that of the clubs in all the other States." So while each team has an ostensible "home" in which it plays and in which its players reside during the season, a baseball team is "primarily an ambulatory organization" — a homeless entity.[121]

(Holmes's father, the poet Oliver Wendell Holmes Sr., claimed to have played baseball at Harvard in the 1820s, or well before Abner Doubleday's fictive "invention" of the sport.)[122]

One consequence of stripping baseball of its antitrust exemption would be, quite probably, an increase in the number of teams, though these would not necessarily be located in currently franchise-less areas. As Mark Rosentraub has speculated, they would likely arise in New York City, Los Angeles, and other major media markets, though as more teams entered the league, the value of a franchise would decline, salaries would fall in response, and so would ticket prices.[123]

Donald Fehr, executive director of the Major League Players Association and one of those attorneys for an unpopular cause (higher salaries for millionaires) that fans love to hate — or at least detest at a level roughly equal to that at which the greedy owners are detested — spilled the owners' dirty little secret in a 1989 interview.

> "Let's be honest," said Fehr. ["Why start now?" the average fan might ask. But I digress.] "The owners have never been interested in expansion, and the nonsense you hear that suggests they are is just that — nonsense.

> The reason is very simple. There's no question there ought to be one, maybe two, teams in Florida. I mean, there's a great baseball tradition there. It's about to be the third-most populous state in the country, and they can't figure out how to put a baseball team anywhere in the state? The reason is if you put a team in Tampa, [Chicago White Sox co-owner Jerry] Reinsdorf can't extort money from the city of Chicago by threatening to move to Tampa.
>
> That's worth more to him than any number of expansion teams are ever going to be worth. And for that, they're not going to expand. To suggest that they're interested in it, they quite simply are not telling the truth. I think it's been a hoax since the beginning."[124]

Strong words and an illuminating perspective, and Fehr was both right and wrong. The owners do keep the numbers of teams artificially low in order to maintain a ready supply of MLB-less cities that can be used to extort money from cities with existing franchises. Consider this observation by Professor George H. Sage of the University of Northern Colorado: "In 1901, the population of the United States was approximately 76 million people and there were 16 major league teams, eight each in the American and National Leagues. From 1901 until 1961, no expansion occurred while the U.S. population more than doubled to 179 million."[125] In 2012, MLB has almost doubled, to a total of 30 teams, though the population of the USA is 311 million. The size of the major leagues has not kept pace with the growth in the American population, to put it mildly. But then that is the point.

Florida did eventually get its two teams: in Miami (which, annoyingly, called itself the Florida Marlins, as if appropriating the whole state) in 1993 and in Tampa (after the White Sox got their stadium deal) in 1998.

The situation in professional football is somewhat different.

Even if NFL owners wanted to block a franchise move, as they certainly did with their nemesis Al Davis of the Los Angeles/Oakland Raiders, they would encounter antitrust troubles. While MLB is exempt from antitrust laws, the NFL is not (except in limited cases), so any attempt by the league to prevent an owner from relocating his team would be met by a lawsuit with an excellent chance of success. The NFL could not stop Davis from jerking the Raiders up and down the California coast, and didn't even try to stop Robert Irsay from moving the Colts from Baltimore to Indianapolis or Art Modell from changing the Cleveland Browns to the Baltimore Ravens. (The Maryland legislature did consider eminent domain proceedings against the Colts — a very bad move, legally and strategically.)

Katherine C. Leone, writing in the *Columbia Law Review*, advocates exempting the NFL from antitrust liability in the matter of franchise relocation regulation, though she would "include a provision that requires the league to reject a team relocation when a city has financed construction or renovation of a stadium and the debt from the improvements has not been retired."[126] Owners would be prohibited from milking a new city's gullible political leadership until the debts incurred by the gullible political leadership of the last city they milked had been paid off. There is a sense in which this is the federal government intervening to protect cities from the stupidity of their own governing class, and whether such interventions make sense is an open question. But as Leone notes, "Football teams that accept public financing for building a stadium are not traditional businesses. While a traditional company may move from a city, such companies do not often require cities to build office buildings for them. By accepting public money, teams have given up some independence and should have an obligation to pay the money back."[127]

Leone goes on to call for greater congressional involvement in deciding which teams may move and which must stay. The factors she suggests are sensible — condition of a team's current stadium and the extent to which it was publicly subsidized, the team's financial position, the existence of local investors who would keep the team put, the good faith shown by the host city in negotiating leases and such — but the broader question is do we really want Congress to "determine the criteria for denying or approving a move," as she recommends.[128] As we have seen — as if we didn't already know it — members of Congress have pressured the major leagues to award franchises to New Orleans and Atlanta (in the NFL) and Denver and Miami (in MLB), among other cities. By an odd coincidence, the influential lawmakers who steered teams into those cities represented those selfsame cities and states. Think of that!

Granting Congress the power to oversee franchise moves, or write the criteria by which the leagues should approve or reject such moves, would politicize the question to a degree that most of us would find excessive. One might as well permit the Senate a say in how many wild card teams make the NFL playoffs, or whether or not to extend the designated hitter rule to the National League.

Of course the easiest way to solve the franchise shortage would be for the NFL (and other major leagues) to admit new teams, but this would greatly impair the ability of rent-seeking (or rent-free seeking) owners from trolling about for subsidies, advertising their available teams to cities like a street corner vendor in stolen merchandise flashing his wares. In, say, a 40 team NFL, owners with blackmail on their minds and thievery in their hearts would not be able to extort a new luxury stadium out of local taxpayers by threatening to move to Los Angeles or Portland or San Antonio or any of the other top candidates on the never-very-secret expansion list. For Los Angeles, Portland, and San Antonio would already have teams and would not be dangling subsidies in front of faithless owners' eyes. Nor would the NFL commissioner be making speeches before local chambers of commerce and hinting that maybe someday, somehow, if the business community showed proper support — that is, lobbied government with enough vigor — an NFL team just might appear in old San Antone.

(Starting up a new league would also expand the number of cities with teams, but despite the artificial scarcity of franchises, rival leagues have generally failed: witness the USFL and the World Football League. The World Hockey Association and the American Basketball Association were at least partially successful, as a handful of their franchises were absorbed into the dominant leagues before they went belly up. But finding a place to play is a real problem. Cities that have prominent teams in the major leagues are unlikely to subsidize the construction of stadiums for startup teams in new leagues. And without the media and publicity that attend to large-city teams, a new league is unlikely to attract the television contracts necessary for survival.)

Leone also suggests that cities could insist upon long-term leases for professional sports teams. But of course signing such a lease would deprive an owner of the ability to troll about for more lucrative subsidies, so that is a serious nonstarter. Besides, in the eyes of too many owners, leases are made to be broken.

The cost to the league of frequent movement can be measured in eroded loyalties, a sense of betrayal, and greater difficulty for fans in identifying with players and teams. The sum of these factors pales, however, when put beside the potential revenue which a city desperate for major league status might offer a faithless owner.

Back the Pack!

For cities wishing to retain sports franchises that are shopping around for better offers, *community ownership* would be a far superior alternative to using the brutal hand of the state via eminent domain. Not *government* ownership, mind you — but *community* ownership.

The model is the Green Bay Packers of the NFL. They have been called "the least-subsidized professional sports team in the country," though it should be noted that this honorific was applied before the Packers gained a subsidy of $251 million toward a $295 million renovation of Lambeau Field in 2003.[129] Nevertheless, despite this bilking of the taxpayers (primarily through a half-cent county sales tax), the Packers are one team that does not routinely hold the locals hostage with threats to move unless they get their way. The Packers, as Lynn Reynolds Hartel explains in the *Loyola of Los Angeles Entertainment Law Journal,* began playing in 1919. In 1923, 3 years after the founding of the NFL, they were organized as a nonprofit corporation under the laws of the State of Wisconsin. Twelve years later, still a nonprofit, they issued 300 shares of common stock. In order for this team located in a city whose population of under 100,000 was by far the smallest in a burgeoning league to survive, the club issued 10,000 shares of common stock, pegged at $25 per share, in 1950. No single owner could possess more than 200 shares. A stock split in 1997 raised $24 million by selling stock at $200 a share to 106,000 buyers. The stock, writes Hartel, "is largely ceremonial because the new shareholders have diluted voting rights, no possibility of profits, and may receive no other special benefits."[130]

So why did Packers' fans buy in? To keep the team healthy, and in Green Bay. With no mercurial owner to threaten movement — a seven-member executive committee and forty-five member board of directors manage affairs — Packers fans aren't tormented by fears of losing their team. True, the franchise could be relocated by a majority vote of shareholders, but with "over ninety percent of the shareholders residing in Green Bay, all presumably rabid Packers fans, it is highly unlikely that any shareholder would ever vote to relocate the team."[131] The "National Floating League," as it bid fair to become when Baltimore, Los Angeles, Oakland, St. Louis, Houston, and Cleveland lost teams in short order, never really threatened Green Bay, even though the Wisconsin city's 100,000 residents could fit snugly into a thin slice of Los Angeles or Houston.

Alas, the Packers' way has been practically foreclosed by the league, which is not eager for other small cities to join what are big-city, big-money clubs. The NFL's Constitution and Bylaws, as Hartel notes, "effectively prohibit community-based corporations from owning an NFL franchise."[132] Hartel has sketched a model under which fans in other cities could band together to support a community-owned team, but the league has shown less than zero interest in permitting such an arrangement. One Green Bay Packers, it would seem, is enough.

As Hartel notes, the NFL does not explicitly ban community-owned teams. The league's bylaws "do not contain any express provisions prohibiting a public corporation from owning a franchise." In theory, as long as the ownership group is for-profit, thousands of fans could band together, buy stock, and own a team that would, in time, become as deeply attached to its home city as the Packers are to Green Bay.

But the devil is in the details. New members of the league are required to provide "names and addresses of all persons who do or shall own any interest or stock in the applicant." These persons are also required to provide written financial statements. Given that Hartel's model for a Packers-type franchise would have as many as 50,000 shareholders, these mandates present certain logistical problems. But they are superable; this requirement is not: "Each proposed owner or holder of any interest in a membership, including stockholders in any corporation…and all other persons holding any interest in the applicant must be individually approved by the affirmative vote of not less than three-fourths or 20, whichever is greater, of the members of the League." As Lynn Reynolds Hartel observes, "gaining League approval of each shareholder would be practically impossible."[133]

The league's ostensible reasons for resisting corporate, or community, ownership of franchises have nothing to do, of course, with the fact that community ownership in particular would anchor teams to cities and reduce the number and the bargaining power of footloose owners shopping franchises to the highest-bidding city council. Former commissioner Pete Rozelle, who successfully opposed New England Patriots' owner William H. Sullivan's attempt (and antitrust suit) to permit him to sell shares of his team to the public, opined that a team having thousands of shareholders would be hamstrung by sheer numbers. Without a single owner, with whom would the NFL deal? How would the team make decisions? Taking it one step further, would the shareholders vote on coaching decisions? On who should play quarterback? (For certain teams — say, the Washington Redskins — turning such matters over to the fans couldn't make things any worse.)

Green Bay is the answer to all these questions. No matter how many tens or even hundreds of thousands of shareholders a franchise might have, an executive committee or board of directors will manage the corporation's affairs, and there is no reason to believe this process will be any more chaotic than it is in Green Bay. Which is to say, not chaotic at all.

The league has expressed other reasons for opposing corporate ownership. These range from a concern about "undue commercialization," à la the Disney Corporation's ownership of the NHL's Anaheim Mighty Ducks, to the desire to keep up the mirage that NFL teams are basically "mom-and-pop operations."[134] The naïvete underlying these two conceptions is positively staggering. The NFL worried about "commercialization"? The league that has turned itself over to television, that permits its games to be slowed and distorted by television timeouts and extended commercials, that marketed such dubious personalities as Bret Favre and Terrell Owens to credulous children — this league frets over commercialization? And what naïf thinks that a league in which men such as Jerry Jones, Dan Snyder, and Paul Allen own teams is "mom and pop"? Even Ari Fleischer's public relations agency would be ashamed to sell a lie like that.

No, the league fears community or corporate-owned teams because it understands that such arrangements would seriously inhibit taxpayer subsidy of its members. As Joseph L. Bast, president of the free-market-friendly Heartland Institute of Chicago, put it, "Allowing fans to own franchises — a model pioneered in 1923 by the Green Bay Packers — would put a stop to the extortionist practice of teams threatening to relocate unless they are subsidized."[135]

Given that Green Bay is the most storied and one of the most popular franchises in the history of the NFL, why would the league close the door to other Green Bays? Lynn Reynolds Hartel suggests that the league's mucka-mucks realize that fan-owned teams would be infinitely less likely to extort subsidies from cities that are desperate for entry into the league.

"Community-owned franchises would effectively shift the burden from taxpayers to individuals interested in owning a piece of an NFL team in a particular community," she writes. Since "the shareholders of [a community-owned] team are highly motivated to keep the team as a permanent fixture in their city," they would not be shopping the team around the country, dangling the prospect of a move and angling for freebies from the city solons of Portland and Las Vegas and San Antonio and all the other NFL-less cities that want in on the action.[136] The more fan- or community-owned teams there are, the slower the gravy train rolls.

Legendary Chicago newspaper columnist Mike Royko saluted his Bears' archrival, the Packers, for their ownership situation:

> If there is one team that truly deserves to be called America's Team, it is the most unlikely community to have a major league sports franchise of any kind. Yes, I'm talking about little Green Bay, Wisconsin, and its Packers.
>
> You don't hear the owners of the Green Bay team whining that they are not rich enough or trying to shake down the local taxpayers for new goodies that will make them even richer.

> That's because the Packer franchise is owned by the kind of people who should own every football franchise.
>
> Basically, it is owned by the people of Green Bay. And it would be almost impossible for the team to go anywhere else because no one individual owns a big enough piece to do it.[137]

Indeed, Green Bay, unlike most other teams in the NFL, has never threatened to relocate.

Few visions are more nightmarish to the NFL than that of cities failing to bid for franchises with taxpayers' money. If a significant number of franchises are tied to their host cities because the shareholders are local fans, the assumed stability of those franchises bodes ill for the other subsidy-seeking owners. And "the NFL does not want to promote non-subsidized privately funded stadiums because it would set a precedent for shifting the economic risk from cities back to the individual owners."[138]

Why, that would be free enterprise. We mustn't have *that!*

There are modest steps the NFL could take if it wanted to replicate the Green Bay experience. Hartel recommends that the reporting and approval requirements for stockholders only apply to those who own, say, 5% of stock. Approving 50,000 shareholders on a case by case basis is impossible; approving 10 or 15 is much more doable. Similarly, a 5% (or even 1%) threshold could be applied when it comes to restrictions such as that barring shareholders from betting on games. Permitting a fan who owns a single share of stock to lay down a fin or sawbuck on the local eleven is not exactly like letting Dan Snyder bet five million on the result of the Cowboys–Redskins game.

Permitting community ownership would hardly be a cure-all to the malady of public sports subsidies. There are publicly owned teams in other sports — the Florida Panthers of the NHL (Sunrise Sports and Entertainment), the Boston Celtics of the NBA (Boston Basketball Partners, LLC), the Toronto Blue Jays of MLB (Rogers Communications) — and that fact has not prevented them from pursuing subsidies. But those are owned by business corporations, not by fans, as in Green Bay. And the general principle enunciated by Hartel stands: "Public ownership," at the very least, can "help shift the burden of financing a team from the general tax-paying public, which currently subsidizes many private NFL owners, to the shareholders of the team." Those who buy the stock will, in essence, be paying for the stadium, the practice field, the league fees, and the player and staff salaries. This would be a throwback to the days when owners bore the risks (and reaped the rewards) of team ownership — and it might, if only in a partial way, "return professional football to its civic foundation."[139]

Where Do We Go from Here?

Our age of increasingly expensive venues in which major league sports are played is not without its slivers of hope for beleaguered taxpayers weary of shelling out for others' amusements.

Mayya M. Komisarchik and Aju J. Fenn note that 56 new venues housing one or more of the four major league sports (NFL, NBA, NHL, and MLB) opened between 1995 and 2009, while at least 14 teams expect to be in new digs by 2015.

The trends for the 56 recently opened arenas and stadiums are not necessarily ominous. The relative percentage of private contributions to construction has increased, although that has been offset by the fact that the overall costs have zoomed upward and the frequency of new construction has greatly increased. The public will bear the burden of two-thirds of the cost of the baseball parks being built between 2010 and 2015, say Komisarchik and Fenn, but the trend in football is striking. "The public financed 84% of NFL venue projects before 1995," they write, "62% of stadium costs incurred between 1995 and 2009, and is scheduled to pay for only 19% of stadiums to be built between 2010 and 2015."[140]

Still, despite the praiseworthy examples of Joe Robbie in Miami, the Bradley Center in Milwaukee, and even, conceding all the complicating factors, the Dodgers in Los Angeles, major-league owners continue to shake down cities and counties for handouts.

With his typical cogency, Neil J. Sullivan lays out the far-reaching implications of the private model: "If modern stadiums and arenas were privately financed, the team would be tied to the community in a way that would preclude frivolous shifts from city to city. What is immoral is not the casual transfer of sports teams but the expenditure of hundreds of millions of public dollars for private entertainment businesses."[141] An owner who has built his own stadium faces powerful disincentives to moving the team: he would be saddled with paying off an empty stadium with no tenant.

So what to do about all this? The answer is obvious, though it will require willpower, fortitude, and integrity on the part of city and state officials — virtues not apparently in overabundant supply among the political class. For which politician wants to be blamed when the Pirates or Jaguars or — God forbid! — Yankees leave town? Arthur T. Johnson urges this course of action: "At whatever costs, local officials must resist threats to relocate. They must be prepared to negotiate hard, and in their public statements show a willingness to allow the team to leave if its price for remaining is too high. Their negotiation style must be the same as with any other business seeking public benefits. The message to the franchise must be that it is welcome to stay, but it is not crucial to the city's future."[142]

Certainly any city official with even the barest knowledge of the literature on the subject would follow such advice. (Surely there are such officials out there!) But as the academic and empirical consensus penetrates even the most resistant domes, the wraith-thin justification of "psychic benefits" is making its appearance. Stadium subsidies will not be vanquished without a bruising battle.

For his part, Mark Rosentraub recommends that "The National Governors' Association and the U.S. Conference of Mayors should establish a binding compact among all its members that prohibits the use of tax dollars and all other revenues collected by governments or publicly created authorities (such as user fees, grants from other governments, lottery proceeds, and so forth) for the building or maintenance of venues used by professional sports teams." He also urges that "no public resources can be used to assemble or provide the land upon which venues used by professional sports teams are built" and that the only public contribution to such teams may be in the provision of public transportation infrastructure — roads, streets, subway lines, etc.[143]

Sounds good, but how to enforce it? The chance of cities banding together to negotiate collectively with team owners, their cooperation serving to counterbalance the monopoly power of the leagues, is slim to none. As Roger G. Noll and Andrew Zimbalist note, "the temptation to cheat by secretly negotiating with a mobile team is too strong to preserve concerted behavior."[144] It is all very well for city officials in, for instance, Baltimore and St. Louis and San Francisco to urge collective bargaining, but why should team-hungry officials in Portland or Columbus throw in with them? In the abstract, even the denser city officials must understand that it would be in their interests to band together to refuse stadium subsidies. But for all their talk of intermunicipal cooperation, the minute a footloose owner winks at potential host cities it's every man, or city, for himself, or itself. And the multitude of indirect subsidies — road and infrastructure improvements, sweetheart leases — would be near impossible to police.

At the end of their long and valuable work, Noll and Zimbalist conclude that reform will not come from the federal government or eminent domain or antitrust law but — just maybe — the fans. "The most likely source of reform, though still a long shot, will be grass roots disgruntlement and citizen education." That is a long shot. More likely, "large-scale public subsidies to wealthy team owners and athletes will be a feature of the professional sports landscape" for years to come.[145] Billionaire owners and millionaire players are in no danger of being tossed off the gravy train any time soon.

Nevertheless, the answer of what is to be done was summarized several years ago by urban analyst Neil R. Peirce, who encapsulated the wise advice of Mark Rosentraub, Associate Dean of the University of Indiana's School of Public and Environmental Affairs. The Peirce-Rosentraub solution? Simple: "Stop subsidizing all professional teams. Make 'em all — baseball, football, hockey, basketball, every one of them — go cold turkey. Refuse to build arenas or ballparks for them. Say 'no' to requests for operating subsidies."[146]

That is much easier said than done. But maybe the sports fans and taxpayers of America should start playing their own brand of hardball.

Notes

1. Neil J. Sullivan, "Major League Baseball and American Cities: A Strategy for Playing the Stadium Game," in *The Economics and Politics of Sports Facilities*, p. 182.
2. Neil deMause and Joanna Cagan, *Field of Schemes: How the Great Stadium Swindle Turns Public Money into Private Profit*, p. xi.
3. Mark S. Rosentraub, *Major League Losers: The Real Cost of Sports and Who's Paying for It* (New York: Basic Books, 1997), pp. 3–4.
4. "Average salary in baseball just over $3 million in 2010," www.CBSSports.com, December 13, 2010.
5. Jarrett Bell, "NFL salaries: Top NFL QBs could be in line for contract hikes," www.usatoday.com, March 9, 2010.
6. Andrew Brandt, "NBA draft picks come up short compared to NFL top rookies," www.huffingtonpost.com, June 24, 2010.
7. Kurt Badenhausen, "The Highest-Paid NHL Players," www.forbes.com, December 1, 2010.
8. "Household Income for States," www.census.gov.
9. Robert A. Baade, "Evaluating Subsidies for Professional Sports in the United States and Europe: A Public-Sector Primer," *Oxford Review of Economic Policy*: 595.
10. Tom Van Riper, "Football's Billionaires," www.forbes.com, October 1, 2010.
11. Robert A. Baade, "Home Field Advantage? Does the Metropolis or Neighborhood Derive Benefit from a Professional Sports Stadium?" in *The Economics and Politics of Sports Facilities*, pp. 85–86.
12. All net worths from Tom Van Riper, "Football's Billionaires," www.forbes.com.
13. Allen R. Sanderson, "Sports Facilities and Development: In Defense of New Sports Stadiums, Ballparks and Arenas," *Marquette Sports Law Journal*, Vol. 10, No. 2 (Spring 2000): 174–75.

14. Dennis Zimmerman, "Subsidizing Stadiums," in *Sports, Jobs & Taxes: The Economic Impact of Sports Teams and Stadiums*, p. 121.
15. Mayya M. Komisarchik and Aju J. Fenn, "Trends in Stadium and Arena Construction, 1995–2015," unpaginated.
16. Ibid.
17. Ibid.
18. Carolyn A. Dehring, Craig A. Depken II, and Michael R. Ward, "The Impact of Stadium Announcements on Residential Property Values: Evidence from a Natural Experiment in Dallas-Fort Worth," Working Paper, 2006, p. 1.
19. Dennis Coates and Brad R. Humphreys, "The Stadium Gambit and Local Economic Development," *Regulation*: 16.
20. James Quirk and Rodney D. Fort, *Pay Dirt: The Business of Professional Sports Teams*, p. xxiii–xxiv.
21. Ibid., p. 131.
22. John Crompton, "Beyond Economic Impact: An Alternative Rationale for the Public Subsidy of Major League Sports Facilities," *Journal of Sport Management*: 41–42.
23. James Quirk and Rodney D. Fort, *Pay Dirt: The Business of Professional Sports Teams*, p. 127.
24. John L. Crompton, Dennis R. Howard, and Turgut Var, "Financing Major League Facilities: Status, Evolution and Conflicting Forces," *Journal of Sport Management*: 168–69.
25. Andrew Moylan, "Stadiums and Subsidies: Home Run for Wealthy Team Owners, Strike-out for Taxpayers," National Taxpayers Union Foundation, NTUF Policy Paper #163, October 30, 2007, p. 3.
26. Ibid., p. 5.
27. Judith Grant Long, "Full Count: The Real Cost of Public Funding for Major League Sports Facilities," *Journal of Sports Economics*, Vol. 6, No. 2 (May 2005): 120.
28. Ibid.: 139.
29. Katherine C. Leone, "No Team, No Peace: Franchise Free Agency in the National Football League," *Columbia Law Review*: n120.
30. Roger G. Noll and Andrew Zimbalist, "Build the Stadium. Create the Jobs!" in *Sports, Jobs & Taxes: The Economic Impact of Sports Teams and Stadiums*, p. 25.
31. John Siegfried and Andew Zimbalist, "The Economics of Sports Facilities and Their Communities," *Journal of Economic Perspectives*: 100.
32. Edward I. Sidlow and Beth M. Henschen, "Building Ballparks: The Public-Policy Dimensions of Keeping the Game in Town," in *The Economics and Politics of Sports Facilities*, p. 165.
33. Kevin J. Delany and Rick Eckstein, "The Devil Is in the Details: Neutralizing Critical Studies of Publicly Constructed Stadiums," *Critical Sociology*, Vol. 29, No. 2 (2003): 203.
34. Dennis Coates and Brad R. Humphreys, "Do Economists Reach a Conclusion on Subsidies for Sports Franchises, Stadiums, and Mega-Events?" *Econ Journal Watch*: 296.
35. Ibid.: 301–302.

36. Mark S. Rosentraub, *Major League Losers: The Real Cost of Sports and Who's Paying for It*, p. 149.
37. Richard Corliss, "Build It, and They (Will) MIGHT Come," *Time*.
38. Alex Williams, "Back to the Future," *New York*.
39. Dennis Coates and Brad R. Humphreys, "Caught Stealing: Debunking the Economic Case for D.C. Baseball," pp. 5–7.
40. Mark S. Rosentraub and David Swindell, "'Just Say No?' The Economic and Political Realities of a Small City's Investment in Minor League Baseball," *Economic Development Quarterly*, Vol. 5, No. 2 (May 1991): 154–55.
41. Steven A. Riess, *Touching Base: Professional Baseball and American Culture in the Progressive Era*, p. 111.
42. Robert A. Baade and Allen R. Sanderson, "The Employment Effect of Teams and Sports Facilities," in *Sports, Jobs & Taxes: The Economic Impact of Sports Teams and Stadiums*, p. 93.
43. Robert A. Baade, "Stadiums, Professional Sports, and Economic Development: Assessing the Reality," pp. 14–15.
44. Ibid., pp. 22–23.
45. Robert A. Baade and Richard F. Dye, "The Impact of Stadiums and Professional Sports on Metropolitan Area Development," *Growth and Change* (Spring 1990): 1.
46. Ibid.: 13.
47. Ibid.: 12.
48. Drake Bennett, "Ballpark figures," *Boston Globe*.
49. Dennis Coates and Brad R. Humphreys, "Do Economists Reach a Conclusion on Subsidies for Sports Franchises, Stadiums, and Mega-Events?" *Econ Journal Watch*: 305.
50. Phillip A. Miller, "The Economic Impact of Sports Stadium Construction: The Case of the Construction Industry in St. Louis, MO," *Journal of Urban Affairs*, Vol. 24, No. 2 (2002): 159.
51. Ibid.: 161.
52. Ibid.: 159.
53. Ibid.: 160, 161.
54. Ibid.: 170.
55. Ian Hudson, "Bright Lights, Big City: Do Professional Sports Teams Increase Employment?" *Journal of Urban Affairs*, Vol. 21, No. 4 (1999): 397–407.
56. Dean V. Baim, *The Sports Stadium as a Municipal Investment*, p. 157.
57. Ibid., pp. 162–63.
58. Robert A. Baade, "Home Field Advantage? Does the Metropolis or Neighborhood Derive Benefit from a Professional Sports Stadium?" in *The Economics and Politics of Sports Facilities*, p. 88.
59. Robert A. Baade, "Stadiums, Professional Sports, and Economic Development: Assessing the Reality," pp. 2, 3.
60. Roger G. Noll and Andrew Zimbalist, "The Economic Impact of Sports Teams and Facilities," in *Sports, Jobs & Taxes: The Economic Impact of Sports Teams and Stadiums*, p. 71.
61. John Siegfried and Andrew Zimbalist, "The Economics of Sports Facilities and Their Communities," *Journal of Economic Perspectives*: 106.

62. Craig A. Depken II, "The Impact of New Stadiums on Professional Baseball Team Finances," *Public Finance and Management*, Vol. 6, No. 3 (June 2006): 436.
63. Ibid.: 436, 446.
64. Carolyn A. Dehring, Craig A. Depken II, and Michael R. Ward, "The Impact of Stadium Announcements on Residential Property Values: Evidence from a Natural Experiment in Dallas-Fort Worth," p. 13.
65. Robert A. Baade, "Home Field Advantage? Does the Metropolis or Neighborhood Derive Benefit from a Professional Sports Stadium?" in *The Economics and Politics of Sports Facilities*, pp. 74, 78.
66. John C. Melaniphy, "The Impact of Stadiums and Arenas," *Real Estate Issues*: 37.
67. Ibid.: 38.
68. Richard Pollard, "Evidence of a reduced home advantage when a team moves to a new stadium," *Journal of Sports Sciences* (2002): 969.
69. Ibid.: 972.
70. *Minor League Baseball and Local Economic Development*, edited by Arthur T. Johnson (Urbana: University of Illinois Press, 1995), p. 32.
71. Ibid., p. 249.
72. Mark S. Rosentraub and David Swindell, "'Just Say No?' The Economic and Political Realities of a Small City's Investment in Minor League Baseball," *Economic Development Quarterly*: 152.
73. Ibid.: 154.
74. Drake Bennett, "Ballpark figures," *Boston Globe*.
75. Mark S. Rosentraub and David Swindell, "'Just Say No?' The Economic and Political Realities of a Small City's Investment in Minor League Baseball," *Economic Development Quarterly*: 164–65.
76. Drew Stone, "Introducing Parkview Field," Fort Wayne *News-Sentinel*, September 12, 2008.
77. Victor A. Matheson and Robert A. Baade, "Have Public Finance Principles Been Shut Out in Financing New Sports Stadiums for the NFL in the United States?" College of the Holy Cross Department of Economics Faculty Research Series, Paper No. 05-11 (July 2005): 2.
78. Ibid.: 14.
79. Ibid.: 5.
80. Mayya M. Komisarchik and Aju J. Fenn, "Trends in Stadium and Arena Construction, 1995–2015."
81. Neil deMause and Joanna Cagan, *Field of Schemes: How the Great Stadium Swindle Turns Public Money into Private Profit*, p. 84.
82. Veronica Z. Kalich, "A Public Choice Perspective on the Subsidization of Private Industry: A Case Study of Three Cities and Three Stadiums," *Journal of Urban Affairs*: 212.
83. Rodney Fort, "Direct Democracy and the Stadium Mess," in *Sports, Jobs & Taxes: The Economic Impact of Sports Teams and Stadiums*, p. 146.
84. Ibid., pp. 161–62.
85. Andrew Zimbalist, "The Economics of Stadiums, Teams and Cities," in *The Economics and Politics of Sports Facilities*, p. 57.

86. Ibid., p. 58.
87. Mayya M. Komisarchik and Aju J. Fenn, "Trends in Stadium and Arena Construction, 1995–2015."
88. Quoted in Neil deMause and Joanna Cagan, *Field of Schemes: How the Great Stadium Swindle Turns Public Money into Private Profit,* pp. 103–104.
89. Allen R. Sanderson, "Sports Facilities and Development: In Defense of New Sports Stadiums, Ballparks and Arenas": 183–84.
90. Herbert J. Rubin, after a series of conversations with economic development officials in the Midwest, described the ravenously credit–hogging philosophy of economic development practitioners as "shoot anything that flies; claim anything that falls." It doesn't matter, really, whether or not the incentives or even outright gifts offered by those public officials charged with economic development were *really* the lures that brought a business to town, or that filled a square with tourists, or that replenished a city's tax coffers: shoot and claim everything and let God (or an especially meticulous researcher) sort 'em out. Herbert J. Rubin, "Shoot Anything that Flies; Claim Anything that Falls: Conversations with Economic Development Practitioners," *Economic Development Quarterly*, Vol. 2, No. 3 (August 1988): 236–51.
91. Steven A. Riess, "Historical Perspectives on Sports and Public Policy," in *The Economics and Politics of Sports Facilities,* p. 33.
92. Joseph L. Bast, "Sports Stadium Madness: Why It Started/How to Stop It," p. 6.
93. William J. Hunter, "Economic Impact Studies: Inaccurate, Misleading, and Unnecessary," Heartland Institute Policy Paper No. 21, July 22, 1988, p. 6.
94. John Crompton, "Beyond Economic Impact: An Alternative Rationale for the Public Subsidy of Major League Sports Facilities," *Journal of Sport Management*: 48.
95. Robert A. Baade, "Stadiums, Professional Sports, and Economic Development: Assessing the Reality," pp. 7–8.
96. Jordan Rappaport and Chad Wilkerson, "What Are the Benefits of Hosting a Major League Sports Franchise?" *Federal Reserve Bank of Kansas City Economic Review* (First Quarter 2001): 59, 63.
97. John Siegfried and Andrew Zimbalist, "The Economics of Sports Facilities and Their Communities," *Journal of Economic Perspectives*: 104.
98. Kevin J. Delany and Rick Eckstein, "The Devil Is in the Details: Neutralizing Critical Studies of Publicly Constructed Stadiums," *Critical Sociology*: 189.
99. Ibid.: 195.
100. Kevin J. Delany and Rick Eckstein, "The Devil Is in the Details: Neutralizing Critical Studies of Publicly Constructed Stadiums," *Critical Sociology*: 197.
101. Ibid.: 197–98.
102. Ibid.: 199.
103. John Siegfried and Andrew Zimbalist, "The Economics of Sports Facilities and Their Communities," *Journal of Economic Perspectives*: 105.
104. Kevin J. Delany and Rick Eckstein, "The Devil Is in the Details: Neutralizing Critical Studies of Publicly Constructed Stadiums," *Critical Sociology*: 206–207.
105. Ibid.: 200.
106. Ibid.: 201.

107. Ibid.: 202, 205, 206, 208.
108. Quoted in Mark S. Rosentraub, *Major League Losers: The Real Cost of Sports and Who's Paying for It*, p. 63.
109. Quoted in John Crompton, "Beyond Economic Impact: An Alternative Rationale for the Public Subsidy of Major League Sports Facilities," *Journal of Sport Management*: 43–44.
110. Mark S. Rosentraub, *Major League Losers: The Real Cost of Sports and Who's Paying for It*, p. 129.
111. Charles C. Euchner, *Playing the Field: Why Sports Teams Move and Cities Fight to Keep Them*, p. 55.
112. Ibid., p. 5.
113. Allen R. Sanderson, "Sports Facilities and Development: In Defense of New Sports Stadiums, Ballparks and Arenas": 190.
114. Robert A. Baade, "Stadiums, Professional Sports, and Economic Development: Assessing the Reality," p. 5.
115. Bruce K. Johnson and John C. Whitehead, "Value of Public Goods from Sports Stadiums: The CVM Approach," *Contemporary Economic Policy*, Vol. 18, No. 1 (January 2000): 52.
116. Jordan Rappaport and Chad Wilkerson, "What Are the Benefits of Hosting a Major League Sports Franchise?" *Federal Reserve Bank of Kansas City Economic Review*: 72.
117. John Crompton, "Beyond Economic Impact: An Alternative Rationale for the Public Subsidy of Major League Sports Facilities," *Journal of Sport Management*: 42.
118. Ibid.: 50.
119. Ibid.: 55.
120. Richard Corliss, "Build It, and They (Will) MIGHT Come," *Time*.
121. *Federal Baseball Club of Baltimore, Inc. v. National League of Professional Baseball Clubs, Et al.*, Supreme Court of the United States, Argued April 19, 1922, Decided May 29, 1922, 259 U.S. 200.
122. Harold Seymour, *Baseball: The People's Game* (New York: Oxford University Press, 1990), p. 131.
123. Neal R. Pierce, "Sports Blackmail: 'Just Say No'?" Cleveland *Plain Dealer*, January 26, 1997.
124. Richard Justice, "Union's Fehr Says Owners Stand Against Expansion," *Washington Post*, March 5, 1989.
125. George H. Sage, "Stealing Home: Political, Economic, and Media Power and a Publicly-Funded Baseball Stadium in Denver," *Journal of Sport and Social Issues*: 112.
126. Katherine C. Leone, "No Team, No Peace: Franchise Free Agency in the National Football League," *Columbia Law Review*: 473.
127. Ibid.: 520.
128. Ibid.: 512.
129. Joseph L. Bast, "Sports Stadium Madness: Why It Started/How to Stop It," p. 40.
130. Lynn Reynolds Hartel, "Community-Based Ownership of a National Football League Franchise: The Answer to Relocation and Taxpayer Financing of NFL Teams," *Loyola of Los Angeles Entertainment Law Journal*, Vol. 18 (1998): 593–94.

131. Ibid.
132. Ibid.: 592.
133. Ibid.: 604–605.
134. Ibid.: 611–12.
135. Joseph L. Bast, "Sports Stadium Madness: Why It Started/How to Stop It," p. 2.
136. Lynn Reynolds Hartel, "Community-Based Ownership of a National Football League Franchise: The Answer to Relocation and Taxpayer Financing of NFL Teams," *Loyola of Los Angeles Entertainment Law Journal*: 602.
137. Quoted in Joseph L. Bast, "Sports Stadium Madness: Why It Started/How to Stop It," p. 41.
138. Lynn Reynolds Hartel, "Community-Based Ownership of a National Football League Franchise: The Answer to Relocation and Taxpayer Financing of NFL Teams," *Loyola of Los Angeles Entertainment Law Journal*: 602.
139. Ibid.: 623, 628.
140. Mayya M. Komisarchik and Aju J. Fenn, "Trends in Stadium and Arena Construction, 1995–2015."
141. Neil J. Sullivan, *The Dodgers Move West*, p. 215.
142. Arthur T. Johnson, "Municipal Administration and the Sports Franchise Relocation Issue," *Public Administration Review*: 526.
143. Mark S. Rosentraub, *Major League Losers: The Real Cost of Sports and Who's Paying for It*, p. 463.
144. Roger G. Noll and Andrew Zimbalist, "Are New Stadiums Worth the Cost?" *Brookings Review* Vol. 15, No. 3 (Summer 1997).
145. Roger G. Noll and Andrew Zimbalist, "Sports, Jobs, and Taxes: The Real Connection," in *Sports, Jobs & Taxes: The Economic Impact of Sports Teams and Stadiums*, pp. 506–507.
146. Neal R. Pierce, "Sports Blackmail: 'Just Say No'?" Cleveland *Plain Dealer*.

About the Author

James T. Bennett is Eminent Scholar and William P. Snavely Professor of Political Economy and Public Policy at George Mason University, and Director of The John M. Olin Institute for Employment Practice and Policy, where he specializes in Public Choice Theory, Political Economy, Labor Economics, and Public Policy. Previously, he served on the faculty of George Washington University. He is the author of dozens of scholarly articles and has served as editor of the *Journal of Labor Research*, on the board of the *Review of Austrian Economics*, and as a referee for such journals as *Economic Inquiry*; *Review of Economics and Statistics*; *Public Choice*; *Public Finance Quarterly*; *Journal of Economics and Business*; and *American Economic Review*. He is the author or editor of more than 20 books, including *From Pathology to Politics* (Transaction), *Unhealthy Charities* (Basic Books), *The Politics of American Feminism* (University Press of America), *Stifling Political Competition* (Springer, SIPC Series), *Not Invited to the Party (Springer)*, and *The Doomsday Lobby* (Springer/Copernicus).

J.T. Bennett, *They Play, You Pay: Why Taxpayers Build Ballparks, Stadiums, and Arenas for Billionaire Owners and Millionaire Players*, DOI 10.1007/978-1-4614-3332-3,

Index

J.T. Bennett, *They Play, You Pay: Why Taxpayers Build Ballparks, Stadiums, and Arenas for Billionaire Owners and Millionaire Players*, DOI 10.1007/978-1-4614-3332-3,

CPSIA information can be obtained at www.ICGtesting.com
Printed in the USA
LVOW10s1521221213

366445LV00003B/145/P